Study Guide for

APPLIED STATISTICS

Study Guide
for

Neter, Wasserman, and Whitmore

APPLIED STATISTICS
Fourth Edition

prepared by

William Sanders
Clarion University

Allyn and Bacon
Boston • London • Toronto • Sydney • Tokyo • Singapore

ISBN 0-205-13483-1

Printed in the United States of America

10 9 8 7 6 5 4 3 2 1   97 96 95 94 93

CONTENTS

PREFACE

HOW TO USE THIS *Study Guide*

Each chapter of the *Study Guide* parallels a chapter in *Applied Statistics*, Fourth Edition. You should study the Text first and then work through the corresponding material in the *Study Guide* because the Text was designed to present new concepts, while the *Study Guide* is designed to reinforce concepts.

Chapters of the *Study Guide* are divided into several sections, each corresponding to one or several sections of the relevant chapter in *Applied Statistics*. For example, Section A in Chapter 1 of the *Study Guide* deals with the material presented in Chapter 1, Sections 1.1 and 1.2 of the Text. The sections of the *Study Guide* are designed so that you should be able to review the corresponding Text material and work through the *Study Guide* in one sitting.

Each section of the *Study Guide*, in turn, is composed of several parts. The first part in nearly all sections is titled "Review of Basic Concepts" and consists of a series of "fill-in-the-blank," "true-false" and similar questions. These review questions are designed so that you can test your understanding of the material covered in the text. Answers for the review questions are found at the end of each chapter of the *Study Guide*. If you have trouble with any of the review questions, you should go back and study that portion of the text in more detail before proceeding to the problem portion of the *Study Guide* section.

In most sections of the *Study Guide*, a problem follows the "Review of Basic Concepts." Problem-solving plays a critical role in the study of statistics, both in understanding key concepts and in being able to apply computational procedures. In solving these problems, you should check your answer against the solution found at the end of the chapter. A correct answer shows that you are on the right track and will reinforce your grasp of the material. Should your answer be incorrect, you will be able to locate the mistake at once from the solution given, and discover for yourself why you made the error. For example, you may find that you do not yet understand some point or concept or, perhaps, are applying a computational procedure incorrectly. If you make a mistake, identify the source of difficulty. Then redo the problem and check your answer again before proceeding to the next section. You will find this approach both time-saving and effective.

You should always work the first problem in each section. On occasion, a second problem follows. If so, this second problem addresses a different concept or procedure than the first one. In such cases, you should continue on with the second problem. There are a few other points which you should note before proceeding. First, a checkmark ($\checkmark$) is used to indicate points where you should compare your answers with those given in the *Study Guide*. Frequent checkmarks are utilized so that you will not go too far in the wrong direction before identifying and correcting the error.

Second, review questions are identified with both a letter and a number (i.e., A.1, A.2,...,B.1, B.2,...) while problems are sequentially numbered (i.e., 1, 2, etc.) through the chapter. This is done to make identification easy since you are often asked to refer back to a previous review question or problem.

Third, you should neatly draw graphs and write calculations on separate sheets of paper and retain these to use when studying for an examination.

Fourth, at times, you may find that your answers differ slightly from those given in the *Study Guide* because of differences due to rounding. These differences need not concern you provided your procedure is correct.

A NOTE TO INSTRUCTORS

Use of this *Study Guide* in conjunction with *Applied Statistics*, Fourth Edition, can make the study of statistics more effective. The review questions provide extensive drill with terminology, notation, new concepts and computational procedures. The problem material is realistic and the settings are varied, designed to motivate the student and to illustrate the usefulness of statistical methods in a variety of administrative situations.

The *Study Guide* is completely self-correcting: answers and/or solutions are provided for all review questions and problems. Thus, correct understanding is immediately reinforced. At the same time, causes of errors are identified so that correct understanding can be achieved.

Material in the *Study Guide* is cross-referenced to *Applied Statistics* section by section rather than chapter by chapter, thereby allowing considerable latitude in topic coverage and assignments.

The *Study Guide* can be incorporated into the course in a number of ways. Some instructors may wish to make formal assignments from the *Study Guide*, with solutions to be handed in. In this case, there would be no need for grading, since answers are given in the *Study Guide*. However, the completed solutions would serve to show that the work has been done. Other instructors may wish to simply leave to the discretion of the student which sections of the *Study Guide* covered in the course are completed. Regardless of which method is utilized, the *Study Guide* can result in more efficient use of classroom time by employing many of the advantages of programmed instruction in reviewing basic concepts and providing practice with solution procedures.

It is possible that some errors remain in the *Study Guide* despite careful checking of the manuscript and proof. I would appreciate having them called to my attention.

I want to give my thanks to the previous edition's author, Kenneth Schneider, on which this edition is based. Also I would like to thank John Neter, G. A. Whitmore, and William Wasserman for their contributions in proofreading this *Study Guide*.

W.V.S.

NOTE TO STUDENTS: It is very important that you understand how to use this Study Guide so that you can derive the benefits it is intended to provide during your study of Applied Statistics. Therefore, if you have not already done so, please read the "How to Use This Study Guide" section of the Preface before proceeding to Section A of this chapter.

A. DATA SETS AND DATA SOURCES
(Text: Chapter 1, Sections 1.1 and 1.2)

Review of Basic Concepts

A.1 A collection of facts or figures which have been gathered for a particular study is called a(n) _____

_____.

A.2 In a statistical study of hospital patients, six characteristics were examined. For each, indicate whether that characteristic has numerical or nonnumerical outcomes.

a. Amount billed to patient. ____   d. Age of patient. _____
b. Reason for hospitalization. ____   e. Patient's race. _____
c. Patient's gender. _____   f. Patient's family size. _____

A.3 For the situation in A.2, is it true that the characteristics having numerical outcomes are called variables while the others are not variables since they have nonnumerical outcomes? Explain.

_____

A.4 For the situation in A.2, hospital records for a large number of patients were used to obtain the desired data. The patients included in this data set are referred to as _____ of the data set, the information on amount billed to any given patient is called a(n) _____ and the information on all six variables for any given patient is called a(n) _____. √

»Problem 1

In a study of employee productivity in the ABC Company, 50 employees of one department were randomly assigned to two groups. Group A was placed on a 4-day, 10-hour workweek, while Group B remained on a 5-day, 8-hour workweek. Partial results of this study follow.

Employee ID Number	Years of Experience	Study Group	Productivity Index	Productivity Change
457-92-47628	13	A	103	No
472-92-49963	4	A	112	Yes
467-77-99326	11	B	93	Yes
.	.	.	.	.
.	.	.	.	.
.	.	.	.	.
396-46-45949	7	B	123	No

a.   What constitute the elements in this data set?   _____

b.   Identify the variables in this data set and indicate whether each is qualitative or quantitative.

Variable	Quantitative or Qualitative
_____	_____
_____	_____
_____	_____
_____	_____

c.   i.   What constitutes the observation on "Years of Experience" for the employee listed first in the preceding table?

   ii.   What constitutes the case or record for this employee?
                                                                      √

d.   Is this a univariate, bivariate or multivariate data set? ____ Explain. _____

e.   Is this data set from an internal or external source? _____ √

B. EXPERIMENTAL AND OBSERVATIONAL STUDIES
(Text: Chapter 1, Section 1.3)

Review of Basic Concepts

B.1 The basic difference between experimental and observational studies is that, in _____ studies, control is exercised over one or more factors to determine their influence on a variable of interest, whereas in _____ studies no such control is exercised.

B.2 In an educational study of the effect of computer training on performance in statistics courses, a statistics class is divided in half at random. One half is given computer training while the other is not. Performance of the two groups is later compared. Is this an experimental or observational study? _____

B.3 In a study of the relationship between gender and voter preference, each person in a sample of registered voters is asked who s/he plans to vote for in an upcoming election. Preferences of male and female voters are later compared. Is this an experimental or observational study? _____

B.4 In comparing experimental or observational studies, it is generally true that a(n) _____ study provides stronger evidence of the effect of one or more factors on the variable of interest. √

»Problem 2

a. Refer to the study of productivity described in Problem 1. Was this an experimental or observational study? _____ Why? _____ √

b. Suppose it is desired to study the impact of employment status of students at your university (employed full-time versus employed part-time versus not employed) on academic performance, as measured by grade point average.

   i.   Explain why it would not be possible to conduct this as an experimental study.

   ii. Describe one approach to conducting this as a non-experimental study. √

C. DATA ACQUISITION AND DEFICIENCIES IN DATA SETS

(Text: Chapter 1, Sections 1.4, 1.5 and 1.6)

Review of Basic Concepts

C.1 Indicate whether each of the following data acquisition procedures involves observation, personal interview, or self-enumeration.

    a.   A survey of shopping habits of area residents conducted over the telephone. _____

    b.   A study of television viewing habits using a device which attaches to a TV set and records when the set is on and the station to which it is tuned. _____

    c.   Distribution of a mail questionnaire to study newspaper reading habits of respondents. _____

    d.   A study of photographic equipment purchases using a warranty registration card. _____

    e.   A study of automobile traffic flow using an automatic traffic counter. _____

C.2 Is the following statement true or false? "Both the averages and limitations of the personal interview arise from the direct contact between the respondent and the interviewer." _____

C.3 Both the advantages and limitations of the self-enumeration procedure arise from the elimination of _____ in the data acquisition process.

C.4 One limitation of self-enumeration is that it can lead to _____ response rates which, in turn, can bias the survey results if those who responded are not _____ of the entire group contacted.

C.5 The question: "Do you subscribe to a Sunday paper? ___ Yes ___ No" is an example of a(n) _____ type of question.

C.6 The question: "Explain in your own words why you do not subscribe to a Sunday paper." is an example of a(n)_____ type of question.

C.7 Is the following statement true or false? "One major disadvantage of free-answer or open-end questions is that they may suggest an answer which the respondent otherwise might not have given." _____

C.8 A computerized data set contains percentage rates of return (ROR) and principal amounts (PRIN) for investments in a large portfolio. How can the return for each investment (RET) be calculated, using logical operations on the computer? How can yet another variable (HIGH) be created to denote the investments which earned above 8%, as opposed to those which earned 8% or less? _____

C.9 Consider the following question. "Don't you agree that viewing a motion-picture version of a book is much better than reading the book, because the motion-picture shows you how each character looks and what she or he is doing?" This is an example of a(n) _____ question, since it suggests a particular answer.

C.10 Identify whether or not each of the following represents an error arising in the data acquisition stage of a study (as opposed to some other kind of error).

     a.   A respondent gave a wrong answer because the interviewer asked a question incorrectly. _____ Yes _____ No

     b.   In a telephone survey, respondents who had changed telephone numbers since the last directory was published could not be reached. _____ Yes _____ No

c. An analyst forgot to include questionnaires returned "in person" in running a computer program to process data. _____ Yes _____ No

d. A traffic counter broke down two hours before the end of a study period and went undetected. _____ Yes _____ No

e. A respondent misstated the number of hours he spent watching TV. _____ Yes _____ No

f. In preparing a final report, a typist reversed the digits of the result "34%," which then appeared in the report as "43%." _____ Yes _____ No √

»Problem 3

Identify a fault in each of the following questions, and reword the question to avoid it.

a. "To the nearest ten dollars, how much money did your family spend dining out in the last twelve months? _____ "

b. "When shopping for boys' jeans, don't you think that the durability is the most important factor? _____ Yes _____ No"

c. "How much do you read? _____ "

d. "What type of work do you do for a living?

_____ Professional   _____ Sales/Clerical
_____ Unskilled Labor   _____ Skilled Labor
_____ Retired"

e. "Did you do a poor, fair or good job at raising your children? _____ " √

Answers to Chapter Reviews and Problems

A.1 data set

A.2 a. numerical   b. nonnumerical   c. nonnumerical

d. numerical   e. nonnumerical   f. numerical

A.3 No; a variable can have either numerical or nonnumerical outcomes. Those with numerical outcomes are called quantitative variables while those with nonnumerical outcomes are called qualitative variables.

A.4 elements; observation, reading or outcome; case or record

Problem 1

a. employees

b. "Years of Experience" is quantitative.
"Study Group" is qualitative.
"Productivity Index" is quantitative.
"Productivity Change" is qualitative.

c. i. "13 years of experience" constitutes this observation.

ii. The information on all four variables (i.e., 13 years of experience, study group A, productivity index of 103 and "no" productivity change) constitutes this case or record.

d. Multivariate; the data set contains four variables.

e. internal

B.1 experimental; observational

B.2 experimental study

B.3 observational study

B.4 experimental study

Problem 2

a.   Experimental; control was exercised over type of workweek.

b.   i.   It would not be feasible to exercise the experimental control required here. Students cannot be randomly assigned one of the three employment status conditions.

ii.   One approach would be to interview a group of students to determine the employment status and grade point average of each. Academic performance could then be compared among various employment status conditions.

C.1 a.   personal interview   b.   observation   c.   self-enumeration

d.   self-enumeration   e.   observation

C.2 true

C.3 interviewers

C.4 low; representative

C.5 multiple-choice

C.6 free-answer or open-end

C.7 false; this is a disadvantage of multiple-choice questions

C.8 RET = ROR*PRIN; IF ROR > 8 THEN HIGH = 1; IF ROR ≤ 8 THEN HIGH = 0

C.9 leading

C.10 a. Yes   b. No   c. No   d. Yes   e. Yes   f. No

Problem 3

a. Respondents probably cannot recall the exact amount. To avoid this fault, one could design a multiple-choice question using ranges of dollar amounts, and ask the respondent to select the most appropriate category. The question also refers to too long a time period. To avoid this fault, one could rephrase the question to ask the amount in the last month rather than the last twelve months.

b. This is a leading question. To avoid this fault, one could design
   the question so it doesn't single out any particular feature. This
   could be done with either a free-answer question or a multiple-choice
   question having a list of possible features.

c. The question is too vague. For example, does the researcher want a
   page count or time estimate; daily weekly or other time frame; all
   material read or just pleasure reading? There are too many possible
   interpretations of this question. To avoid this fault, the question
   must be rewritten with considerably greater clarity.

d. The list of choices is incomplete. For example, homemakers who are
   not in the labor force and unemployed persons are omitted. To avoid
   this fault, the list of choices must be expanded, possible even to
   include a choice titled "Other - Specify _____."

e. This is a "prestige question." Few parents will admit they did a
   poor job of raising their children. (Incidentally, if you are not
   satisfied with your reworded question, don't be discouraged. This is
   an example of a question which is extremely difficult, if not
   impossible, to ask directly without getting biased responses.)

A. QUANTITATIVE DATA
(Text: Chapter 2, Section 2.1)

Review of Basic Concepts

A.1 A listing of the observations on a quantitative variable in increasing or decreasing order of magnitude is called a(n) _____.

A.2 The ages of six automatic presses in a plant are 10, 6, 8, 3, 6, and 11 years. Place these observations in an array.

_____

A.3 In a stem-and-leaf display with two-digit stems and one-digit leaves, the outcome 136 is to be recorded. For this outcome, the digits _____ are recorded on the stem while the digit _____ is recorded on the leaf.

A.4 Nine households in a study of fifty households reported annual incomes (measured to the nearest $ thousand) between $10 thousand and $19 thousand, inclusive. These incomes were 14, 10, 17, 17, 19, 12, 13, 18, and 19. If the digit 1 (corresponding to tens of thousands of dollars) constitutes a stem of the stem-and-leaf display for this data set, the stem appears as _____

_____.

A.5 When the number of elements in each class of a frequency distribution is expressed as a percent of the total number of elements of the data set, the distribution is called a(n) _____ _____ distribution.

A.6 What problem is encountered if too many classes are used in constructing a frequency distribution? _____

_____

A.7 What problem is encountered if too few classes are used in constructing a frequency distribution? _____

_____

A.8 When constructing a frequency distribution, is it generally better to have equal or unequal class sizes? _____

A.9 A class in a frequency distribution which appears as either the first or last class and has only one limit, upper or lower, is referred to as a(n) _____ class.

A.10 A bar graph of a frequency distribution is called a(n) _____.

A.11 What is a frequency polygon? _____

_____

A.12 The percent frequency distribution of minutes required by 50 workers to complete an assembly task is shown in the following table. A histogram Is to be constructed using a unit of width of 10 minutes. Therefore, the heights of the histogram bars are to be

_____ , _____ , _____ ,

and _____ , respectively.

Minutes	Percent of Workers
20 - under 40	4.0
40 - under 60	18.0
60 - under 70	36.0
70 - under 80	42.0
	100.0

A.13 For the situation in A.12, a less-than cumulative frequency distribution is to be constructed. The cumulative percent frequency for "less than 60 minutes" is _____ percent. Interpret this percent. _____

_____

A.14 The graph of a cumulative frequency distribution is called a(n) _____ . √

»**Problem 1**

a. An analyst is studying pricing policies in a large supermarket. As part of the study, a sample of 60 food items priced below $1.00 has been selected. Unit prices of these 60 items appear in Table 2.2.

Table 2.1
SUPERMARKET PRICES
FOR 60 ITEMS PRICED BELOW $1.00
(See Problem 1a)

Item	Unit Price	Item	Unit Price	Item	Unit Price	Item	Unit Price
1	.65	16	.88	31	.25	46	.90
2	.39	17	.79	32	.47	47	.55
3	.89	18	.39	33	.75	48	.84
4	.74	19	.42	34	.19	49	.39
5	.75	20	.33	35	.70	50	.40
6	.45	21	.25	36	.89	51	.49
7	.29	22	.84	37	.43	52	.61
8	.36	23	.45	38	.79	53	.34
9	.85	24	.79	39	.76	54	.85
10	.56	25	.81	40	.99	55	.58
11	.95	26	.52	41	.92	56	.39
12	.89	27	.35	42	.65	57	.73
13	.46	28	.38	43	.59	58	.48
14	.35	29	.49	44	.67	59	.75
15	.70	30	.21	45	.45	60	.79

   i. Construct a stem-and-leaf display of these unit prices.√

  ii. Within what range(s) of values are these unit prices concentrated? _____

 iii. Looking at specific prices, does there appear to be any special concentrations of final digits in the data set? _____ Explain. _____
_____

  iv. Construct a back-to-back stem-and-leaf display of prices from the main aisles (items 1-30) and prices from peripheral displays (items 31-60).

   v. Construct a cumulative distribution of the first 15 items. √

b.   Following is the net return on investment last year (in percent)
        for each of 60 life insurance companies.

        5.85   6.18   6.30   6.08   5.26   5.30   4.18   4.20   5.71   5.92
        6.04   5.81   6.14   6.02   5.87   3.96   5.92   5.66   5.27   4.34
        3.75   6.39   5.04   4.06   6.12   5.96   4.55   4.72   4.99   5.62
        6.14   5.67   5.01   5.58   6.18   5.24   4.98   6.01   4.22   6.18
        4.94   5.96   5.70   5.98   4.84   5.74   6.02   4.46   6.33   4.30
        4.98   6.54   6.37   5.85   5.96   5.04   5.28   5.08   5.44   4.64

   i.   In part (1) of the following table, construct a frequency
            diagram and percent frequency diagram for this data set, using
            the given classes.√

  ii.   In part (2) of the table, construct a less-than cumulative
            percent frequency distribution for this data set. √

	(1) Frequency Distribution		(2) Cumulative Frequency Distribution	
Net Return on Investment	Number of Companies	Percent of Companies	Less Than This Net Return	Cumulative Percent of Companies
3.50-under 4.00	_____	_____	3.50	
4.00-under 4.50	_____	_____	4.00	_____
4.50-under 5.00	_____	_____	4.50	_____
5.00-under 5.50	_____	_____	5.00	_____
5.50-under 6.00	_____	_____	5.50	_____
6.00-under 6.50	_____	_____	6.00	_____
6.50-under 7.00	_____	_____	6.50	_____
			7.00	_____
Total	_____	(_____)		

 iii.   Using separate graphs, construct a histogram and frequency
            polygon for the percent frequency distribution constructed in
            part (b.i.).

  iv.   Construct an ogive corresponding to the less-than cumulative
            percent frequency distribution which was constructed in part
            (b.ii.).√

B. QUALITATIVE DATA

(Text: Chapter 2, Section 2.3)

Review of Basic Concepts

B.1 The classes in any system of data classification must be
_____ and _____. What does this requirement
mean? _____

B.2 What is the basic objective in determining classes to be used in a
qualitative distribution? _____

B.3 A graphic presentation of a qualitative distribution often takes
the form of a(n) _____ _____.

B.4 Bars in a bar chart typically differ in _____ but not in
_____.

B.5 When constructing a bar chart of more than one qualitative
distribution, the distributions are usually expressed in
_____ terms to help in comparing the distributions. √

»Problem 2

The Pollution Control Agency of a large metropolitan area received
complaints about noncompliance with pollution control regulations
against 65 companies last year. Forty companies were in the urban area
while the remaining 25 were in suburban areas. The PCA kept a record of
the industry classification of each company subject of a complaint.
Table 2.1 contains a list of the 40 urban and 25 suburban companies and
their industrial classifications, presented in the order in which the
complaints were received. The industries used in the classification
were (1) manufacturing (manufac), (2) retail/wholesale (ret/whol), (3)
transportation (trans), (4) service, and (5) other.

Table 2.2
COMPLAINTS ABOUT NONCOMPLIANCE WITH POLLUTION
CONTROL REGULATIONS (See Problem 2)

(a) Urban Companies

Company	Industry	Company	Industry	Company	Industry	Company	Industry
1	Manufac	11	Manufac	21	Other	31	Manufac
2	Service	12	Trans	22	Manufac	32	Ret/Whol
3	Ret/Whol	13	Ret/Whol	23	Ret/Whol	33	Service
4	Manufac	14	Manufac	24	Trans	34	Manufac
5	Trans	15	Other	25	Manufac	35	Ret/Whol
6	Other	16	Manufac	26	Service	36	Service
7	Ret/Whol	17	Ret/Whol	27	Manufac	37	Trans
8	Manufac	18	Service	28	Trans	38	Ret/Whol
9	Trans	19	Manufac	29	Other	39	Manufac
10	Service	20	Trans	30	Trans	40	Manufac

(b) Suburban Companies

Company	Industry	Company	Industry	Company	Industry	Company	Industry
1	Trans	7	Ret/Whol	13	Manufac	19	Other
2	Ret/Whol	8	Service	14	Ret/Whol	20	Manufac
3	Service	9	Trans	15	Service	21	Service
4	Other	10	Service	16	Ret/Whol	22	Manufac
5	Manufac	11	Trans	17	Trans	23	Trans
6	Service	12	Other	18	Service	24	Service
						25	Ret/Whol

a.i. Complete the following qualitative distribution for the 40
urban companies, using the indicated classes. √

Industry	Number of Companies	Percent of Companies
Manufacturing	_____	_____
Retail/Wholesale	_____	_____
Transportation	_____	_____
Service	_____	_____
Other	_____	_____
Total	_____	(_____)

ii. Construct a bar chart depicting the qualitative distribution in part (i). Use the number of companies in constructing your chart.√

b.   i. Construct a bar chart to compare the distributions of urban and suburban companies by industry classification.

ii. Comment on the major differences between the distributions of urban and suburban companies depicted on your chart in part (i). _____

_____

_____ √

C.   BIVARIATE AND MULTIVARIATE DATA
(Text: Chapter 2, Sections 2.4 and 2.5)

Review of Basic Concepts

C.1 A sample of 2,000 adults is to be classified simultaneously by gender and by whether or not each adult is a regular cigarette smoker. The resulting cross-classification is called a(n) _____   _____ distribution.

C.2 For the section in C.1, each adult is classified into _____ of _____ possible cells of the distribution.

C.3 Three hundred personal computers of a certain model are to be classified simultaneously by age (0 to 2 years, 3 to 5 years, 6 to 10 years) and by number of breakdowns to date (0, 1, 2, 3 or more). The resulting cross-tabulation is called a(n) _____ _____ distribution. This distribution contains _____ classification cells.

C.4 A(n) _____ is used to depict the relationship between two quantitative variables graphically. √

»**Problem 3**

The Admissions and Records Department of a small midwestern university has conducted a study of students who have dropped out of the university. Among other variables, there students have been classified by gender and class (e.g., freshman, sophomore). The Admissions and Records Department discovered that 50 students failed to return to the university for the Spring Semester. Gender and class for these students are presented in Table 2.3.

Table 2.3
GENDER AND CLASS OF STUDENTS
NOT RETURNING TO THE UNIVERSITY
(see Problem 3)

Student	Gender	Class	Student	Gender	Class
1	male	freshman	26	male	sophomore
2	female	senior	27	male	junior
3	male	senior	28	male	senior
4	male	sophomore	29	female	junior
5	female	junior	30	male	sophomore
6	female	freshman	31	female	junior
7	male	freshman	32	male	sophomore
8	male	senior	33	male	senior
9	male	sophomore	34	male	freshman
10	female	senior	35	male	sophomore
11	female	junior	36	female	freshman
12	male	junior	37	male	junior
13	female	senior	38	female	senior
14	male	senior	39	male	freshman
15	female	freshman	40	male	sophomore
16	male	senior	41	female	junior
17	female	sophomore	42	female	sophomore
18	male	senior	43	male	junior
19	female	sophomore	44	female	senior
20	male	junior	45	male	junior
21	female	freshman	46	female	freshman
22	male	junior	47	male	senior
23	male	sophomore	48	male	junior
24	female	freshman	49	female	senior
25	female	junior	50	female	junior

a.   Construct a bivariate qualitative distribution for students who didn't return for Spring Semester cross-classified by gender and class. √

b. Convert this bivariate qualitative distribution into percent form. Include percent of row, percent of column and percent of total for each cell in the distribution. √

c. Use the results in parts (a.) and (b.) to determine the following:

   i. How many students that dropped out were sophomores and female? _____

   ii. What percent of the 28 males were juniors? _____
  iii. What percent of the freshman were females? _____
   iv. What percent of the 50 students were sophomores? _____

   v. What percent of the 50 students were male juniors? _____ √

D. DATA MODELS AND RESIDUAL ANALYSIS
(Text: Chapter 2, Section 2.6)

Review of Basic Concepts

D.1 The difference between an observed value and corresponding anticipated value is called a(n) _____.

D.2 Based on past experience, it is known that 40 percent of all campers at a popular state park stay one night. In a recent sample of 200 campers, 73 stayed one night. In this sample, the observed value for the number of campers staying one night is _____ _____ and the anticipated value is _____.

D.3 For the situation in D.2, the residual for the number of campers staying one night is _____. √

»**Problem 4**

Recall the situation in Problem 1(a). The analyst here is studying pricing policy in a large supermarket. A stem-and-leaf display of unit prices of 60 food items priced below $1.00 in the supermarket was constructed in Problem 1(2.i.). One price policy that is frequently used in retailing is to determine the unit price by adding a constant percentage amount for overhead and profit onto the wholesale price of the item. Such a price policy is called cost-plus or markup pricing. Under this model, one would anticipate that final digits in unit prices would occur with equal frequency.

    a.   If cost-plus pricing is used by the supermarket, about how often should each digit appear in the leaves of the stem-and-leaf display? _____ √

    b.   Construct a residual plot for the final digits in the leaves of the stem-and-leaf display.

    c.   Based on your residual plot, does it appear as if the supermarket is using cost-plus pricing? _____ Explain. _____

»**Problem 5**

A data set for fifteen outlets of a chain of music shops is given below. All fifteen shops have similar characteristics, except for the prices charged for compact disks (CDs) and the location of the shop. The data set includes a store number (variable Shop), the monthly dollar value of sales or revenue (variable REV, $ thousand), the monthly unit sales of compact disks (variable CD), and a location variable (M for shops in malls, O for others). Using the symbols associated with the location variable, M and O, construct a multivariate plot that can be used to explore the relationship between unit sales and dollar revenue at CD shops.

Table 2.4
CD Sales Data

Shop	CD	REV	Location	Shop	CD	REV	Location	Shop	CD	REV	Location
1	4	7	O	6	10	16	M	11	10	13	O
2	22	13	M	7	12	11	O	12	25	6	O
3	25	10	M	8	4	12	M	13	19	13	M
4	16	11	O	9	17	16	M	14	7	10	O
5	9	15	M	10	12	18	M	15	19	9	O

Answers to Chapter Reviews and Problems

A.1 array

A.2 In increasing order of magnitude, the array is: 3, 6, 6, 8, 10, 11.

A.3 13; 6

A.4 1|0 2 3 4 7 7 8 9 9

A.5 percent frequency

A.6 The classification does not effectively summarize the data.

A.7 The information is so condensed that it leaves little insight into the pattern of the distribution.

A.8 equal

A.9 open-end

A.10 histogram

A.11 A frequency polygon is a line graph of a frequency distribution.

A.12 $4.0/2 = 2.0$; $18.0/2 = 9.0$; $36.0/1 = 36.0$; $42.0/1 = 42.0$

A.13 22.0; 22% of the workers completed the assembly in less than 60 minutes.

A.14 ogive

Problem 1

a.i.iv.

Unit Prices (in cents)											Aisles		Periphery					

```
   Unit Prices (in cents)                Aisles      Periphery
1 ║ 9                                            9 ║ 1
2 ║ 1 5 5 9                                      5 ║ 2 ║ 1 5 9
3 ║ 3 4 5 5 6 8 9 9 9 9                      9 9 4 ║ 3 ║ 3 5 5 6 8 9 9
4 ║ 0 2 3 5 5 5 6 7 8 9 9            9 8 7 5 3 0 ║ 4 ║ 2 5 5 6 9
5 ║ 2 5 6 8 9                            9 8 5 ║ 5 ║ 2 6
6 ║ 1 5 5 7                              7 5 1 ║ 6 ║ 5
7 ║ 0 0 3 4 5 5 5 6 9 9 9 9      9 6 5 5 3 0 ║ 7 ║ 0 4 5 9 9 9
8 ║ 1 4 4 5 5 8 9 9 9              9 9 5 4 ║ 8 ║ 1 4 5 8 9
9 ║ 0 2 5 9                            9 2 0 ║ 9 ║ 5
```

(a) Stem-and-Leaf Display   (b) Back-to-back Display

Figure 2.1

ii. Unit prices are concentrated in the range $0.33 to $0.49 and also in the range $0.70 to $0.89.

iii. Yes; there appears to be a disproportionately large number of unit prices which end in the digits 5 and 9.

v.

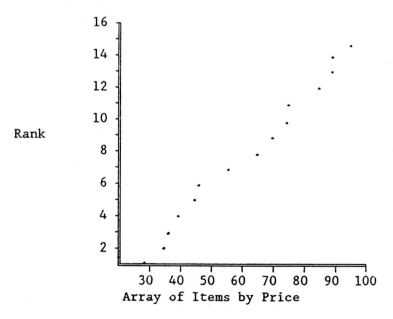

Figure 2.2

b. i., ii.

Net Return on Investment	Number of Companies	Percent of Companies				Less Than This Net Return		Cumulative Percent of Companies		
3.50 - under 4.00	2	(2/60)(100)	=	3.3		3.50				0.0
4.00 - under 4.50	6	(6/60)(100)	=	10.0		4.00	0.0 + 3.3 =	3.3		
4.50 - under 5.00	9	(9/60)(100)	=	15.0		4.50	3.3 + 10.0 =	13.3		
5.00 - under 5.50	9	(9/60)(100)	=	15.0		5.00	13.3 + 15.0 =	28.3		
5.50 - under 6.00	18	(18/60)(100)	=	30.0		5.50	28.3 + 15.0 =	43.3		
6.00 - under 6.50	15	(15/60)(100)	=	25.0		6.00	43.3 + 30.0 =	73.3		
6.50 - under 7.00	1	(1/60)(100)	=	1.7		6.50	73.3 + 25.0 =	98.3		
Total	60			100.0		7.00	98.3 + 1.7 = 100.0			

iii.

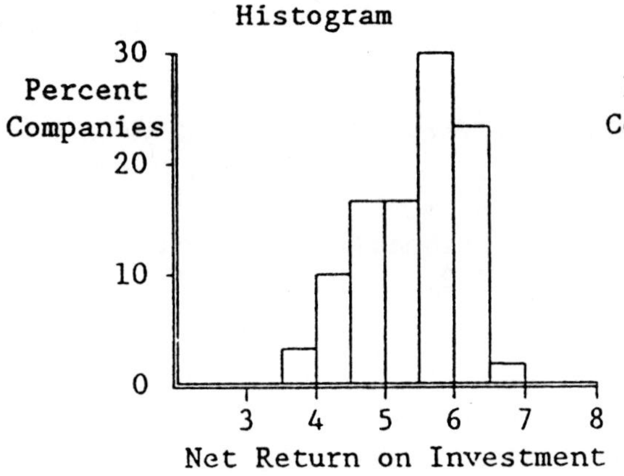

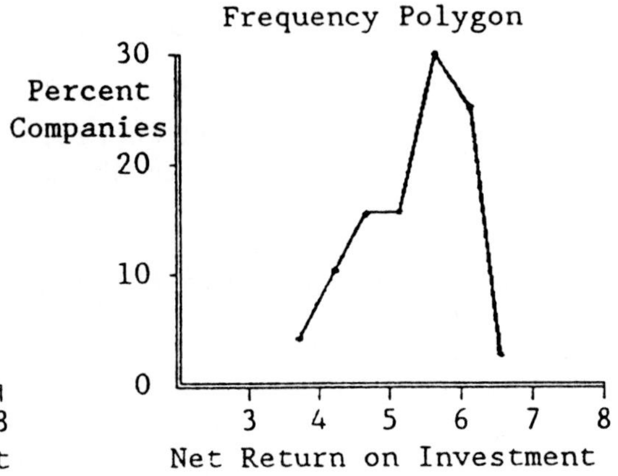

Figure 2.3

(Note: The histogram has bars which span the width of each class and the frequency polygon has class frequencies plotted corresponding to the class midpoints.)

iv.

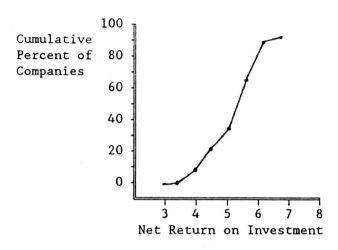

Figure 2.4

B.1 Mutually exclusive; exhaustive. This means that every element in the data set must fall into one and only one class of the system.

B.2 The basic objective is to select a classification system that loses as little essential information as possible, while still effectively summarizing the data.

B.3 bar chart

B.4 length; width

B.5 percent

Problem 2

a. i.

Industry	Number of Companies	Percent of Companies
Manufacturing	14	(14/40)(100) = 35.0
Retail/Wholesale	8	(8/40)(100) = 20.0
Transportation	8	(8/40)(100) = 20.0
Service	6	(8/40)(100) = 15.0
Other	4	(4/40)(100) = 10.0
Total	40	100.0 (40)

ii.

INDUSTRY

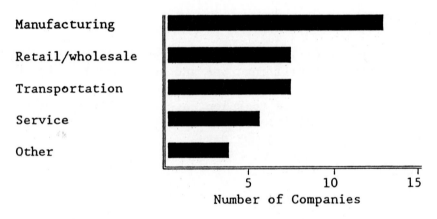

Figure 2.5

b.   i. To compare the different qualitative distributions, we
        construct bar charts in percent terms.

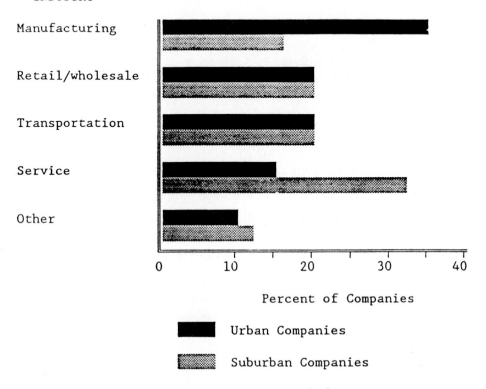

Figure 2.6

ii.   The percents for urban and suburban companies are very similar
        for the retail/wholesale, transportation, and "other" industry
        classifications. However, a relatively higher percent of the
        urban companies were in the manufacturing industry while a
        relatively higher percent of the suburban companies were in
        the service industry.

C.1 bivariate qualitative

C.2 1; 4 (i.e., 2 classes for gender times 2 classes for smoking
    status)

C.3 bivariate frequency; 12 (i.e., 3 classes for age times 4 classes
    for number of breakdowns)

C.4 scatterplot

Problem 3

a.

		Class			
Gender	Freshman	Sophomore	Junior	Senior	Total
Male	4	8	8	8	28
Female	6	3	7	6	22
Total	10	11	15	14	50

b.   To convert this bivariate qualitative distribution into
       percent form, we proceed as follows:

First, each cell entry is expressed as a percent of its row total.
Thus, the class "male; freshman" is $(4/28)(100) = 14.3\%$ of the
total number of males. These percents appear first in the cells of
the following table.

Second, each cell entry is expressed as a percent of its column
total. Thus, the class "male; freshman" is $(4/10)(100) = 40.0\%$ of
the total number of freshmen. These percents appear next in the
cells of the table.

Third, each cell entry is expressed as a percent of the total for
the whole distribution. Thus, the class "male; freshman" is
$(4/50)(100) = 8.0\%$ of the total number of students who failed to
return to the university. These percents appear last in the cells
of the table.

Finally, each row or column total is expressed as a percent of the
total for the whole distribution. Thus, the 10 freshmen comprise
$(10/50)(100) = 20.0\%$ of the total number of students who failed to
return to the university.

| | Class | | | | |
Gender	Freshman	Sophomore	Junior	Senior	Total
Male	14.3 40.0 8.0	28.6 72.7 16.0	28.6 53.3 16.0	28.6 57.1 16.0	56.0
Female	27.3 60.0 12.0	13.6 27.3 6.0	31.8 46.7 14.0	27.3 42.9 12.0	44.0
Total 1	20.0	22.0	30.0	28.0	100.0

  c.   i. 3
    ii. 28.6
   iii. 60.0
    iv. 22.0
    v. 16.0

D.1 residual

D.2 73; .40(200) = 80

D.3 73-80 = -7

Problem 4

a.   With 60 items and 10 digits, each final digit should appear about six times.

b.   The residual for each final digit is obtained as the difference between the observed frequency of that digit in the data set and the anticipated frequency of six. For example, since the final digit 0 appears four times in the data set, the corresponding residual is 4 - 6 = -2. Residuals for other final digits are similarly obtained, yielding the following residual plot.

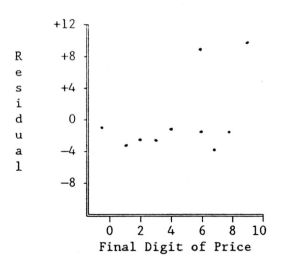

Figure 2.7

c.   No; at least not exclusively. The large positive residuals
       associated with the final digits 5 and 9 indicate that some
       price policy other than (or in addition to) cost-plus pricing
       is in effect. Further investigation may show, for example,
       that unit prices in some food categories are rounded to the
       nearest $.05 or $.09, perhaps to give the impression of a
       "bargain" price.

Problem 5

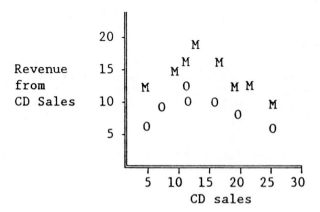

Figure 2.8

DATA SUMMARY MEASURES

NOTE TO STUDENTS: Beginning with Chapter 3 of your Text, a certain amount of mathematical terminology is used. If you are unfamiliar with this notation or need a brief review, it would be advisable to read "Appendix A - Mathematical Review" in your Text prior to working Chapter 3 of this Study Guide.

A. MEASURES OF POSITION - MEAN
(Text: Chapter 3, Section 3.2)

Review of Basic Concepts

A.1 The mean of a set of values is computed by _____ the values and dividing this result by the number of items. Symbolically, this computation is written X = _____ .

A.2 The six nurses assigned to Station 12 of a hospital have 15, 7, 7, 5, 15 and 2 years of experience, respectively. The mean of this set of values is _____ .

A.3 For the situation in A.2, in what units is the mean expressed? _____ .

A.4 The difference between any one value in a data set and the mean of all the values in the data set can be represented symbolically as _____ . The sum of all these differences for a data set equals _____ .

A.5 If A is the value for which the expression $\sum_{i=1}^{n}(X_i - A)^2$ is a minimum, then A must be equal to the _____ of the values $X_1, X_2, \ldots, X_n$.

A.6 An instructor wishes to compute the overall mean score on an exam. Two sections of students took the exam. The first section of 20 students had a mean score of 66 and the second section of 30 students had a mean score of 71. If (3.5) in your Text is used to compute a weighted average X, then f_1 = _____ , f_2 = _____ , X_1 = _____ , X_2 = _____ , and n = _____ .

A.7 For the situation in A.6, $\overline{X}$ = _____ .

A.8 The condition "approximately equal" is denoted by the symbol _____ .

A.9 Consider the following array: 1, 2, 3, 5, 6, 6, 9, 56. In this situation, $\overline{x}$ is strongly influenced by the value _____ , which is referred to as a(n) _____ _____ .

A.10 For the situation in A.10, the arithmetic mean is _____ while the 50 percent trimmed mean is _____ . √

»Problem 1

a.   United Way charitable contributions (in $ thousands) from seven districts were 4, 8, 4, 1, 9, 12 and 4. What is the mean of this data set? _____ √

b.   The mean prices per hundred weight that a rancher received for his cattle in the last three years were $29.00, $23.00 and $25.00. The quantities sold in these years were 4,000, 10,000 and 8,000 hundred weight, respectively. Compute the mean price per hundred weight received by the rancher for the three years combined. _____ √

B.   MEASURES OF POSITION - MEDIAN, MODE, AND PERCENTILES

(Text: Chapter 3, Section 3.2)

Review of Basic Concepts

B.1 The median of a set of items is the value of the _____ item in an array of the items.

B.2 The median tends to be more indicative of typical values in a data set than the mean if the data set contains one or more _____ .

B.3 The median of a data set is denoted by _____ .

B.4 Consider the following array: 14, 31, 38, 42, 43, 48, 49. In this situation the median is the value of the item with rank _____ . Hence, Md = _____ .

B.5 Consider the following array: 6, 8, 8, 8, 8, 8, 8, 8, 9, 9, 10. The median of this set of items is Md = _____ . For this data set, is it accurate to say that about one-half of the items are smaller than the median and about one-half are larger? _____ _____ Explain. _____

B.6 The median can be viewed as the _____ percentile of the data set.

B.7 The 25th percentile of a data set is also called the _____ _____ . √

»Problem 2

The ages of six ice cream vendors at a ballpark are 16, 21, 18, 62, 16 and 24.

   a.   Compute the median age of the vendors. _____

   b.   Construct a step-function cumulative percent frequency distribution for this data set.

   c.   Using the step-function cumulative percent frequency distribution in (c.), find the 15th and 75th percentiles. √

C.   MEASURES OF VARIABILITY
(Text: Chapter 3, Section 3.3)

Review of Basic Concepts

C.1 The range is the difference between the _____ value and _____ value in a set of items.

C.2 The interquartile range is the difference between the
_____ and the _____
_____ of a set of items.

C.3 Is the range or interquartile range more strongly influenced by the
presence of extreme values in a data set? _____
Explain. _____

C.4 Formula (3.12) in your Text is the formula for calculating the
variance of a data set. Indicate the symbolic notation for each of
the following steps in this calculation.

   a.   Compute the difference between each value in the data set and
        the mean of the data set. _____

   b.   Square each of these differences. _____

   c.   Sum these squared differences. _____

   d.   Divide the sum of the squared differences by one less than the
        number of items in the data set. _____

C.5 In a data set giving years of job experience of business
executives, $s^2 = 19.76$. In what units is this measure expressed?
_____

C.6 The positive square root of the variance is called the
_____ _____ and is denoted by the
symbol _____. In what units is this measure expressed
for the data set in C.5?

C.7 Two data sets have standard deviations of 8.4 weeks and 3.9 weeks,
respectively. Which data set has the greater variability as
measured by the standard deviation? _____

C.8 A data set consists of the values 8, 8, 8, and 8. The standard
deviation of this data set, by inspection, must equal
_____.

C.9 The standard deviation measures _____ variability
while the coefficient of variation measures _____
variability.

C.10 The coefficient of variation is equal to the _____

_____ divided by the _____ ,

expressed as a percent, and is denoted by the symbol

_____ . √

»Problem 3

A social worker recorded the following times (in days) to process six
family support claims: 1, 6, 4, 4, 12, 3. (The mean of this data set
is 5.0.)

a.   Calculate the range of these six values. _____
      In what units is the range expressed? _____ √

b.   Compute the variance of this data set using (3.12) in your
      Text. _____

c.   Compute the variance of this data set using (3.12a) in your
      Text. _____ √

d.   Obtain the standard deviation of this data set. _____
      In what units is the standard deviation expressed? _____
      _____

e.   Obtain the coefficient of variation of this data set. _____
      In what units is the coefficient of variation expressed?
      _____

D. SUMMARY MEASURES - OTHER ISSUES
(Text: Chapter 3, Sections 3.4 to 3.6)

Review of Basic Concepts

D.1 A unimodal frequency distribution in which the mean, median and
     mode are equal is called a(n) _____ distribution.

D.2 Among the mean, median and mode, the _____ tends to be located furthest out toward the tail of a skewed unimodal distribution.

D.3 A unimodal frequency distribution which has a tail to the left is skewed _____.

D.4 If a unimodal frequency distribution is skewed positively, the mean will generally be larger than the _____ which, in turn, will generally be larger than the _____.

D.5 A measure of skewness that is obtained by averaging the cubed deviations about the mean is called the _____ _____ about the mean, and is denoted by _____.

D.6 Consider a data set consisting of the following values: 4, 10, 10, 6, 10. The mean of this data set is _____, the second moment about the mean (i.e., the variance) is _____, and the third moment about the mean is _____.

D.7 The standardized skewness measure, denoted by _____, is obtained by dividing the third moment about the mean by the _____ of the standard deviation.

D.8 If a frequency distribution is skewed to the right, then m_3 and m_3' will be _____ _____ zero. If a frequency distribution is symmetrical, then m_3 and m_3' will be _____ _____ zero.

D.9 For a particular data set, $m_3 = -72.0$. Hence the data set is skewed _____. Given the standard deviation for this data set is $s = 3.0$, the standardized skewness measure m_3' is _____.

D.10 The mean and standard deviation of a data set are, respectively, $\overline{X} = 80$ and $s = 6$. The standardized value corresponding to the observation $X_9 = 95$ is $Y_9 =$ _____.

D.11 For the situation in D.10, if all the observations in this data set are expressed in standardized values, the mean and standardized values will be _____.

D.12 A study was undertaken to compare the total construction times (in weeks) of a sample of 400 new homes started in the spring or summer and 150 new homes started in the fall or winter in a northern region. Based on the appropriate descriptive statistics, the following box plots were obtained.

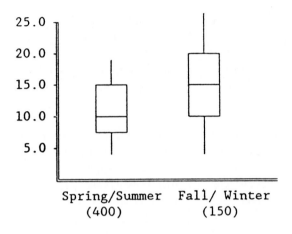

Figure 3.1

Compare the two data sets on the basis of these box plots.

√

E. CALCULATING SUMMARY MEASURES FOR FREQUENCY DISTRIBUTIONS
(Text: Chapter 3, Section 3.7)

Review of Basic Concepts

E.1 Explain each symbol in the formula for approximating the mean of a data set presented as a frequency distribution: $X = \dfrac{\sum\limits_{i=1}^{k} f_i M_i}{n}$

a. k _____   b. f_i _____

c. M_i _____   d. n _____

E.2 Explain each symbol in the formula for approximating the median of a n_1 data set presented as a frequency distribution:

$$Md \approx L + \frac{n_1}{n_2} I.$$

a. L _____   c. n_1 _____
b. n_2 _____   d. I _____

E.3 In a frequency distribution with equal class intervals, the class with the largest frequency is called the _____

_____ .

E.4 There are two essential differences between calculating the variance of a data set and approximating it from a frequency distribution. One is that, in the approximation, $\bar{x}$ is subtracted not from each value in the data set but from the _____ of each class. The other is that each squared difference $(M_i - \bar{x})^2$ is multiplied by the _____ of its class before being summed.

   »Problem 4

Table 3.1 contains a frequency distribution of the net return on invest last year (in percent) for 60 life insurance companies.

Table 3.1
FREQUENCY DISTRIBUTION OF
NET RETURN ON INVESTMENT FOR
60 LIFE INSURANCE COMPANIES
(See Problem 1c.)

Net Return on Investment	(1) i	(2) Cumulative frequency	(3) M_i	(4) $f_i M_i$
3.50 - under 4.00	2	2	3.75	7.50
4.00 - under 4.50	6	8	4.25	.50
4.50 - under 5.00	9	17	___	___
5.00 - under 5.50	9	26	___	___
5.50 - under 6.00	18	44	___	___
6.00 - under 6.50	15	59	___	___
6.50 - under 7.00	1	60	___	___
Total	60			

a. Complete the computations in Columns 3 and 4 of Table 3.1.

b. Approximate the mean net return on investment using this frequency distribution. _____ √

»Problem 5

Consider the frequency distribution in Table 3.1.

a. Approximate the median net return on investment using this frequency distribution. √

b. Approximate the 30th percentile using this frequency distribution. _____

c. Determine the modal class of this frequency distribution.
_____ √

»Problem 6

Table 3.2 contains a frequency distribution of clothing expenditures last year for a sample of 100 middle-class families. (The approximate value of $\overline{X}$ based on the frequency distribution in Table 3.2 is $800.)

Table 3.2
FREQUENCY DISTRIBUTION OF CLOTHING EXPENDITURES
LAST YEAR FOR 100 FAMILIES
(See Problem 6b.)

Expenditures	(1) f_i	(2) Cumulative frequency	(3) M_i	(4) $M_i - \overline{X}$	(5) $(M_i - \overline{X})^2$	(6) $f_i(M_i - \overline{X})^2$
$0 - under $200	2	2	100	-700	490,000	980.000
$200-under $400	4	6	300	-500	250,000	1,000,000
$400-under $600	11	17	500	_____	_____	_____
$600-under $800	28	45	700	_____	_____	_____
$800-under $1000	35	80	900	_____	_____	_____
$1000-under $1200	20	100	1,100	_____	_____	_____
Total	100					

a. What is an upper limit on the range of the expenditure data from which this frequency distribution was constructed? _____ Why is this not the range but an upper limit on the range? _____

b. The approximate first quartile (25th percentile) of this data set is $657.14. Approximate the third quartile and determine the interquartile range. _____ _____ √

c. Complete the computations in columns 4, 5 and 6 of Table 3.2 and obtain an approximation for the variance of clothing expenditures. _____ √

d. Compute the standard deviation from your result in part (iii). _____ √

e. Compute the coefficient of variation from the preceding results. _____

Answers to Chapter Reviews and Problems

A.1   summing; $\overline{X} = \sum\limits_{i=1}^{n} X_i/n$

A.2 $\dfrac{15+7+7+5+15+2}{6} = \dfrac{51}{6} = 8.5$

A.3 The mean is expressed in number of years of experience (i.e., 8.5 years), the same units in which the characteristic of the data set is expressed.

A.4 $X_i - \overline{X}$; zero

A.5   mean

A.6 $f_1 = 20$; $f_2 = 30$; $\overline{X}_1 = 66$; $\overline{X}_2 = 71$; $n = 50$

A.7 $\dfrac{20(66)+30(71)}{50} = 69.0$

A.8 $\approx$

A.9   56; extreme value

A.10 11; $(3+5+6+6)/4 = 20/4 = 5.0$

Problem 1

    a. $\overline{X} = (4+8+4+1+9+12+4)/7 = \$6.0$ thousand

    b. $\overline{X} = \dfrac{4000(29)+10000(23)+8000(25)}{4000+10000+8000} = \dfrac{546,000}{22,000} = \$24.818$

B.1   middle

B.2   extreme values

B.3   Md

B.4   4; 42

B.5 8; No; This interpretation of the median is not accurate for data sets in which the median value is repeated extensively. In this data set, only $1/11 \approx 9.1$ percent of the values are smaller than the median and only $3/11 \approx 27.3$ percent are larger.

B.6 50th

B.7 first quartile

Problem 2

a.

Value:	16	16	18	21	24	62
Rank:	1	2	3	4	5	6

Md = (18 + 21)/2=   19.5 years

b. and c.

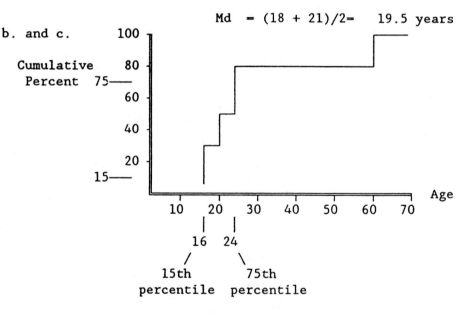

Figure 3.2

C.1 largest; smallest

C.2 third quartile or 75th percentile; first quartile or 25th percentile

C.3 Range; because it depends on the largest and smallest values while the interquartile range is determined only by the central half of the data.

C.4 a. $(X_i-\overline{X})$

b. $(X_i-\overline{X})^2$

c. $\sum\limits_{i=1}^{n}(X_i-\overline{X})^2$

d. $\sum\limits_{i=1}^{n}(X_i-\overline{X})^2/(n-1)$

C.5 years of experience squared

C.6 standard deviation; s; years of experience

C.7 The first data set has greater variability since its standard
deviation is higher

C.8 zero

C.9 absolute; relative

C.10 standard deviation; mean; C

Problem 3

a. $12 - 1 = 11$; days

b. $s^2 = [(1-5)^2+ (6-5)^2+ (4-5)^2+ (4-5)^2+ (12-5)^2+ (3-5)^2]/(6-1)$

$= [(-4)^2+ (1)^2+ (-1)^2+ (-1)^2+ (7)^2+ (-2)^2]/5 = 14.40$

c. $s^2 = \dfrac{1^2+6^2+4^2+4^2+12^2+3^2-[(1+6+4+4+12+3)^2/6]}{6-1}$

$= \dfrac{222 - [(30)^2/6]}{5} = (222 - 150)/5 = 14.40$

d. $s = \sqrt{14.40} = 3.795$; days

e. $C = 100(3.795/5.0) = 75.9$; percent

D.1 symmetrical

D.2 mean

D.3 negatively

D.4 median; mode

D.5 third moment; m_3

D.6 $\bar{X} = 8.0$; $m_2 = s^2 = 8.0$; $m_3 = [(4-8)^3 + (10-8)^3 + (10-8)^3 + (6-8)^3]/(5-1) = [-64+8+8-8+8]/4 = -48/4 = -12.0$

D.7 m_3'; cube

D.8 greater than; equal to

D.9 negatively (to the left); $m_3' = -72/(3)^3 = -2.67$

D.10 $Y_9 = (95.0 - 80.0)/6.0 = 2.50$

D.11 $35.0 - 3(2.5) = 27.5$; $35.0 + 3(2.5) = 42.5$

D.12 Based on the box plots, it appears that new homes started in the fall or winter typically take longer to construct than those started in the spring or summer (from the overall location of the plots as well as their respective medians). Further, construction times are more variable among homes started in the fall or winter (from the respective ranges and interquartile ranges). Finally, construction times for homes started in the spring and summer tend to be more positively skewed (from the lengths of the whiskers as well as locations of the medians within the boxes).

E.1 a.   k is the number of classes in a frequency distribution.

   b. f_i is the frequency of the ith class.

   c. M_i is the midpoint of the ith class.

   d.   n is the total number of items.

E.2 a. L is the lower limit of (median) class containing the middle item.

b. n_1 is the number of items that must be covered in the median class in order to reach the middle item.

c. n_2 is the frequency of the median class.

d. I is the width of the median class.

E.3 modal class

E.4 midpoint; frequency

Problem 4

Net Return on Investment	(1) f_i	(3) M_i	(4) f_iM_i
3.50 - under 4.00	2	3.75	7.50
4.00 - under 4.50	6	4.25	25.50
4.50 - under 5.00	9	4.75	42.75
5.00 - under 5.50	9	5.25	47.25
5.50 - under 6.00	18	5.75	103.50
6.00 - under 6.50	15	6.25	93.75
6.50 - under 7.00	1	6.75	6.75
Total	60		327.00

b. $\bar{X} \approx 327/60 = 5.450$ percent

Problem 5

a.

Net Return On Investment	f_i	Cumulative Frequency
3.50 - under 4.00	2	2
4.00 - under 4.50	6	8
4.50 - under 5.00	9	17
5.00 - under 5.50	9	26
5.50 - under 6.00	18	44*
6.00 - under 6.50	15	59
6.50 - under 7.00	1	60
Total	60	

* Class containing 30th item

The middle value of the distribution occurs at rank 60/2=30. Thus, the median is in the class 5.50-under 6.00, with lower limit L=5.50 and width I=0.50. To reach the median, n_1=4 items (i.e., the 27th, 28th, 29th and 30th items) if the n_2=18 items in the class must be covered. Hence, Md ≈ 5.50 + (4/18)0.50 = 5.611 percent.

b. The 30th percentile corresponds to the cumulative frequency .30(60) = 18. Thus the 30th percentile is in the interval 5.00-under 5.50. Hence, we obtain the following: 30th percentile ≈ 5.00 + (1/9)0.50 = 5.056 percent.

c. The modal class is 5.50 - under 6.00.

Problem 6

a. Upper limit on range is 1,200 - 0 = $1,200. This is an upper limit since we do not know the exact value of the largest and smallest values in the data set. We only know that the largest must be less than $ 1,200 and the smallest cannot be less than $0.

b. 75th percentile ≈ 800.00+(30/35)200.00 = $971.43. Therefore, the interquartile range is approximately 971.43 - 657.14 = $314.29.

c.

Expenditures	(1) f_i	(3) M_i	(4) $M_i-\overline{X}$	(5) $(M_i-\overline{X})^2$	(6) $f_i(M_i-\overline{X})^2$
$0 - under $ 200	2	100	−700	490,000	980,000
$200 - under $ 400	4	300	−500	250,000	1,000,000
$400 - under $ 600	11	500	−300	90,000	990,000
$600 - under $ 800	28	700	−100	10,000	280,000
$800 - under $1000	35	900	100	10,000	350,000
$1000 - under $1200	20	1100	300	90,000	1,800,000
Total	100				5,400,000

$s^2 \approx (5,400,000)/99 = 54,545$

d. $s = -1/54,545 = \$233.55$

e. $C = 100(233.55/800) = 29.194\%$

A. RANDOM TRIALS, SAMPLE SPACES AND EVENTS
(Text: Chapter 4, Section 4.1)

Review of Basic Concepts

A.1 There are two basic components to the definition of a random trial. First, it must be an activity with _____ or more possible outcomes and, second, there must be _____ regarding which outcome will occur.

A.2 The different possible outcomes of a random trial are called _____.

A.3 The set of all possible basic outcomes of a random trial is called the _____ _____.

A.4 The number of basic outcomes of a random trial is denoted by the symbol _____. The ith basic outcome is denoted by _____.

A.5 A sample space consists of four basic outcomes. Symbolically, this sample space is represented as _____.

A.6 A sample space and its _____ _____ are equivalent in structure to a system of _____ and its classes.

A.7 A sample space is called _____ if the basic outcomes of a random trial refer to one characteristic, _____ if they refer to ·two characteristics, and _____ if they refer to three or more characteristics.

A.8 A graphic device that can be used to visualize sample spaces is a(n) _____ _____.

A.9 A subset of basic outcomes of a sample space is called a(n) _____.

A.10 The set of all basic outcomes not contained in event E is called the _____ event to E and is denoted by _____.

A.11 Event F consists of the first basic outcome in a sample space containing four basic outcomes. Symbolically, event F consists of _____ and event F* consists of _____.

A.12 Event G has outcomes o_1 and o_2, and event H has outcomes o_2 and o_3. The set of all basic outcomes (o_1, o_2 and o_3) is called the _____ of G and H, and is denoted by _____.

A.13 For the situation in A.12, the basic outcome o_2 is called the _____ of G and H, and is denoted by _____.

A.14 Event L has outcomes o_1, o_2 and o_4, and event M has outcome o_3. Events L and M have _____ basic outcome(s) in common. Hence, they are _____ _____ events, which can be denoted symbolically as L $\cap$ M = _____. √

»Problem 1

a.   In a market survey, a homemaker is asked to select which one of three different package designs (A, B, C) she finds most attractive. Develop a sample space for this random trial. √

b.   In a study of stock prices, a randomly selected stock is classified by two characteristics; where the stock is traded (national or local) and the price change since the last trading day (increase, decrease, or unchanged).

   i.   Present this bivariate sample space in tabular form. √

   ii.   Present the sample space in a tree diagram. √

c.   A sample space consists of possible winter driving conditions on a Minnesota freeway and contains the basic outcomes $o_1 \equiv$ excellent, $o_2 \equiv$ good, $o_3 \equiv$ fair, $o_4 \equiv$ poor, and $o_5 \equiv$ closed. Consider the following three events.

   E $\equiv$ clear roadway, which includes o_1 and o_2.

   F $\equiv$ hazardous roadway, which includes o_4 and o_5.

G $\equiv$ passable roadway, which includes o_1, o_2, o_3 and o_4.

i.   Which pairs of events, if any, are mutually exclusive?
_____

ii.   Determine the set of basic outcomes for each of the
following and describe in words the driving condition
implied by each.

G* = _____

F $\cap$ G = _____

E U F = _____

F U G = _____

B. PROBABILITY POSTULATES AND PROBABILITY DISTRIBUTIONS
(Text: Chapter 4, Sections 4.2 and 4.3)

Review of Basic Concepts

B.1 The probability of any basic outcome or event can be as low as
_____ or as high as _____.   The probabilities
associated with basic outcome o_i or event E are denoted
symbolically by _____ and _____, respectively.

B.2 The probability of event E can be obtained by _____
probabilities of the _____ _____ contained
in E.   This is denoted symbolically as P(E) = _____.

B.3 The sum of the probabilities of all basic outcomes in a sample
space is _____.

B.4 If events E_1 and E_2 are mutually exclusive, $P(E_1 \cap E_2)$ =
_____.

B.5 The _____ interpretation of probability is based upon
the relative frequency of occurrence in the long run under constant
causal conditions while the _____ interpretation rests
upon degree of personal belief.

B.6 The _____ interpretation of probability can be applied
to any event while the _____ interpretation is limited
just to repeatable events.

B.7 A soccer fan is overheard saying that the odds of his favorite team
winning an upcoming match is 7 to 5. Given this odds ratio, the
assessed probability of that team winning is _____ .

B.8 The probability that a particular machine will malfunction sometime
during the work week is .05. Thus, the odds in favor of a
malfunction are _____ to _____ .

B.9 The basic outcomes in a bivariate sample space are determined by
the intersection of events concerning the two variables which
define the sample space. The probability of any one such basic
outcome is called a(n) _____ probability and the set of all
such probabilities forms a(n) _____ .

B.10 The probability distribution that results from summing joint probabilities either across each row or down each column is called a(n)
_____ probability distribution.

B.11 The probability of an event E_1 occurring given that event E_2 occurs
is called a(n) _____ probability and can be
computed as $P(E_1 | E_2)$ = _____ .

B.12 Consider the following joint probability distribution.

	B_1	B_2	B_3
A_1:	05	.10	.25
A_2:	.10	.15	___

   Determine each of the following probabilities.

   a. $P(A_2 \cap B_3)$= _____ ; $P(A_1)$= _____ ; $P(A_2)$= _____

   b. $P(B_1)$= _____ ; $P(B_2)$= _____ ; $P(B_3)$= _____

   c. $P(B_1 | A_1)$= _____ ; $P(B_2 | A_1)$= _____ ; $P(B_3 | A_1)$= _____

   d. $P(A_1 | B_3)$= _____ ; $P(A_2 | B_3)$= _____ . ✓

»Problem 2

The number of replacement parts of a particular type which will be required by each of two machines during the next production period will be either 1 or 2. The sample space for this situation contains nine basic outcomes, where Ai represents the number of parts required by machine #1 and Bj represents the number of parts required by machine #2. The joint probability distribution for this problem is as follows.

	Parts Required by Machine #2			
Parts Required by Machine #1	0 B_1	1 B_2	2 B_3	Total
A_1: 0	.30	.14	.06	_____
A_2: 1	.18	.10	.02	_____
A_3: 2	.12	.06	.02	_____
Total	___	___	___	_____

a. Obtain the respective marginal probability distributions for parts required by machine #1 and by machine #2. √

b. Obtain the conditional probabilities for number of parts required by machine #2 (Bj) given that no parts are required by machine #1 (A1), $P(B_1|A_1)$ = _____; $P(B_2|A_1)$= _____; $P(B_3|A_1)$= _____ √

c. Construct the conditional probability distribution for number of parts required by machine #1 given two parts are required by machine #2. √

C. BASIC PROBABILITY THEOREMS
(Text: Chapter 4 , Section 4.4)

Review of Basic Concepts

C.1 The theorem which states that the probability of event E can be computed as 1.0 minus the probability of event E* is the _____ Theorem.

C.2 The theorem which can be used to compute the probability of the union of two events, $P(E_1 \cup E_2)$, is called the _____ Theorem. According to this theorem, $P(E_1 \cup E_2) =$ _____.

C.3 If events E_1 and E_2 are mutually exclusive, the Addition Theorem reduces to $P(E_1 \cup E_2) =$ _____.

C.4 $P(E) = .45$; $P(F) = .60$ and $P(E \cap F) = .40$. Thus, $P(E \cup F) =$ _____.

C.5 The probability of event M occurring is .2. The probability of event N occurring given M occurs is .05. With this information, the joint probability of events M and N occurring, which is denoted by _____, can be obtained with the _____ Theorem.

C.6 For the situation in C.5, $P(M \cap N) =$ _____.

C.7 There is a .05 probability that a patient has a certain disease (event A_1). Given that a patient has the disease, a lab procedure has a .97 probability of a positive result (event B_1 given event A_1) and a .03 probability of a negative result (event B_2 given event A_1). Thus, the probability that a patient has the disease and the lab result is positive is _____.

C.8 For the situation in C.7, the probability that a patient has the disease and the lab result is negative is _____. ✓

»Problem 3

Refer to the joint probability distribution for the number of replacement parts required by each of two machines, which appears in Problem 2. Consider the following events for this situation.

E = no replacement part is needed for machine #1.
F = the same number of replacement parts is needed for each machine.
G = a total of three or more replacement parts is needed for the two machines combined.

a.   Obtain each of the following probabilities.

   $P(E) \equiv$ _____ ;
   $P(F) \equiv$ _____ ;
   $P(G) \equiv$ _____ ;
   $P(E \cap F) \equiv$ _____ ;
   $P(E \cap G) \equiv$ _____ ;
   $P(F \cap G) \equiv$ _____ . √

b.   Compute each of the following probabilities using the Addition
   Theorem.
   $P(E \cup F) \equiv$ _____ ;
   $P(E \cup G) \equiv$ _____ ;
   $P(F \cup G) \equiv$ _____ .

c.   Compute each of the following probabilities using the
   Complementation Theorem.
   $P(E^*) \equiv$ _____ ;
   $P(F^*) \equiv$ _____ ;
   $P(G^*) \equiv$ _____ . √

d.   Let event H = scheduled production for the next period is met
   and event H* = scheduled production for the next period is not
   met. Using relative frequencies calculated from observations
   in the recent past, the production manager has determined that
   $P(H|G) = .73$ and $P(H*|G) = .27$, where G is the event defined
   earlier. Compute and interpret each of the following joint
   probabilities.

   i. $P(G \cap H) =$ _____ .

   ii. $P(G \cap H*) =$ _____ . √

D. STATISTICAL INDEPENDENCE AND DEPENDENCE
(Text: Chapter 4, Section 4.5)

Review of Basic Concepts

D.1 If two events are statistically independent, their joint
   probability is equal to the _____ of their marginal
   probabilities. Expressing this symbolically, E_1 and E_2 are
   statistically independent if $P(E_1 \cap E_2) =$ _____ .

D.2 $P(M) = .05$ and $P(N) = .70$. Events M and N are statistically independent. Hence, $P(M \cap N) =$ _____.

D.3 $P(E) = .3$, $P(F) = .45$ and $P(E \cap F) = .15$. Are events E and F statistically independent or statistically dependent? _____. Explain. _____

D.4 Two variables are statistically independent if all the _____ probabilities are equal to the product of their corresponding _____ probabilities.

D.5 The following probability distribution shows the marginal probabilities for two variables. Suppose it is known that these two variables are statistically independent. Determine the appropriate joint probabilities.

	Variable B		
Variable A	B_1	B_2	Total
A_1	———	———	.20
A_2	———	———	.70
A_3	———	———	.10
Total	.80	.20	1.00

D.8 For the situation in D.5, suppose instead it is known that these two variables are statistically dependent. Would the appropriate joint probabilities then differ from those determined in D.5? _____ Explain. _____ √

»**Problem 4**

Based on past experience, an architectural design firm with corporate affiliations to a general contractor (GC) has constructed the following joint probability distribution for the variables how a project culminates (constructed by affiliated GC, constructed by a different GC, not constructed) and a project cost estimate (under $1 millíon, $1 million or more).

Preliminary Cost Estimate

Project Culmination	Under $1 Million B_1	$1 Million or More B_2	Total
A_1 : Affiliated GC	.04	.26	.30
A_2 : Different GC	.08	.02	.10
A_3 : Not Constructed	.48	.12	.60
Total	.60	.40	1.00

a.   i.   Which, if any, pairs of events comprising joint outcomes in this probability distribution are statistically independent? _____

_____

  ii.   Based on your results in part 1i.), what conclusion can be reached about the statistical relationship between how a project culminates and the preliminary project cost estimate? _____
_____ √

b.   i.   Complete the following conditional probability distributions for how a project culminates given the preliminary cost estimate.

Conditional on
Preliminary Cost Estimate

Project Culmination	Under $1 Million B_1	$1 Million or More B_2	Total
A_1 : Affiliated GC	.04/.60=.067	_____	.30
A_2 : Different GC	_____	_____	.10
A_3 : Not Constructed	_____	_____	.60
Total	1.000	1.000	1.000

  ii.   Using these conditional distributions, describe the nature of the statistical relationship between the two variables. _____
_____ √

Answers to Chapter Reviews and Problems

A.1 two; uncertainty

A.2 basic outcomes

A.3 sample space

A.4 k; o_i

A.5 Sample space is comprised of basic events o_1, o_2, o_3 and o_4.

A.6 basic outcomes; classification

A.7 univariate; bivariate; multivariate

A.8 tree diagram

A.9 event

A.10 complementary; E*

A.11 Event F consists of o_1; F* consists of outcomes o_2, o_3, and o_4.

A.12 union; G U H

A.13 intersection; G ∩ H

A.14 no; mutually exclusive; L ∩ M = ϕ

Problem 1

a.   The sample space for this problem is not uniquely defined. If
     the purpose of the study is to examine preference among the
     three designs, the sample space might consist of the three
     basic outcomes o_1 - homemaker selects design A, o_2 - homemaker
     selects design B and o_3 - homemaker selects design C. However,
     if the purpose of the study is to examine the preference for
     design C, the sample space might consist of the two basic
     outcomes o_1 - homemaker selects design C and o_2 - homemaker
     does not select design C. Still other sample spaces might be
     defined, depending on the purpose of the study.

b.   i.

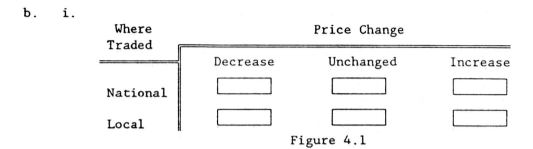

Figure 4.1

ii.

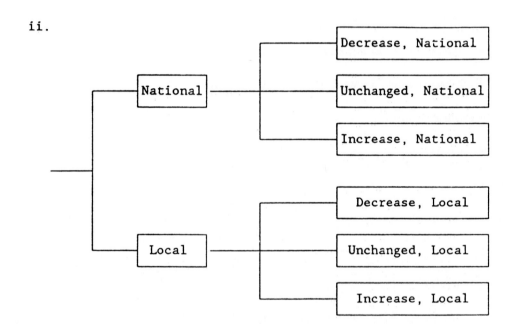

c.   i.   E and F are mutually exclusive since they have no basic
          outcomes in common. The remaining pairs, E and G, F and
          G, are not mutually exclusive.

     ii.   G* is comprised of basic outcome o_5, not passable
          (closed); F ∩ G is o_4 (hazardous and passable, i.e. poor
          driving condition); E U F is o_1, o_2, o_4, or o_5 (anything
          but fair condition); F U G is o_1, o_2, o_3, o_4, o_5 (any
          condition).

B.1 0; 1; $P(o_i)$; $P(E)$

B.2 summing; basic outcomes; $P(E) = \sum\limits_{E} P(o_i)$

B.3 1.0

B.4 $P(E_1 \cap E_2) = 0$

B.5 objective; subjective

B.6 subjective; objective

B.7 $7/(7 + 5) = .583$

B.8 .05 to .95, or 1 to 19

B.9 joint; joint probability distribution

B.10 marginal

B.11 conditional; $P(E_1|E_2) = \dfrac{P(E_1 \cap E_2)}{P(E_2)}$

B.12 a. $P(A_2 \cap B_3) = -35$; $P(A_1) = .40$; $P(A_2) = .60$

b. $P(B_1) = .15$; $P(B_2) = .25$; $P(B_3) = .60$

c. $P(B_1|A_1) = .05/.40 = .125$; $P(B_2|A_1) = .10/.40 = .250$;
$P(B_3|A_1) = 25/.40 = .625$

d. $P(A_1|B_3) = .25/.60 = .417$; $P(A_2|B_3) = .35/.60 = .583$

Problem 2

a.

Ai	P(Ai)	Bj	P(Bj)
A_1 .30+.14+.06 = .50		B_1 .30+.18+.12 = .60	
A_2 .18+.10+.02 = .30		B_2 .14+.10+.06 = .30	
A_3 .12+.06+.02 = .20		B_3 .06+.20+.02 = .10	
Total	1.00	Total	1.00

b. $P(B_1|A_1) = .30/.50 = .60$; $P(B_2|A_1) = .14/.50 = .28$;

$P(B_3|A_1) = .06/.50 = .12$

c.

Parts Required by Machine #1	Conditional Probability Given B_3
0	.06/.10= .60
1	.02/.10= .20
2	.02/.10= .20
Total	1.00

C.1 Complementation

C.2 Addition; $P(E_1 \cup E_2) = P(E_1) + P(E_2) - P(E_1 \cap E_2)$

C.3 $P(E_1 \cup E_2) = P(E_1) + P(E_2)$

C.4 $P(E \cup F) = .45 + .60 - .40 = .65$

C.5 $P(M \cap N)$; Multiplication

C.6 $P(M \cap N) = P(M)P(N|M) = .20(.05) = .01$

C.7 $P(A_1 \cap B_1) = P(A_1)P(B_1|A_1) = .05(.97) = .0485$

C.8 $P(A_1 \cap B_2) = P(A_1)P(B_2|A_1) = .05(.03) = .0015$

Problem 3

a. $P(E) = P(A_1 \cap B_1) + P(A_1 \cap B_2) + P(A_1 \cap B_3) = .50$;
   $P(F) = P(A_1 \cap B_1) + P(A_2 \cap B_2) + P(A_3 \cap B_3) = .42$;
   $P(G) = P(A_3 \cap B_2) + P(A_2 \cap B_3) + P(A_3 \cap B_3) = .10$;
   $P(E \cap F) = P(A_1 \cap B_1) = .30$;
   $P(E \cap G) = 0$ (since E and G are mutually exclusive);
   $P(E \cap G) = P(A_3 \cap B_3) = .02$.

b. $P(E \cup F) = .50 + .42 - .30 = .62$;
   $P(E \cup G) = .50 + .10 = .60$;
   $P(E \cup G) = .42 + .10 - .02 = .50$.

c. $P(E*) = 1 - .50 = .50$;
      $P(F*) = 1 - .42 = .58$;
      $P(G*) = 1 - .10 = .90$.

d.   i. $P(G \cap H) = P(G)P(H|G) = .10(.73) = .073$ is the probability
            that three or more replacement parts are needed and the
            scheduled production is met.

      ii. $P(G \cap H*) = P(G)P(H*|G) = .10(.27) = .027$ is the
            probability that three or more replacement parts are
            needed and the scheduled production is not met.

D.1 product; $P(E_1 \cap E_2) = P(E_1)P(E_2)$

D.2 $P(M \cap N) = .05(.70) = .035$

D.3 statistically dependent; since $P(E \cap F) = .15$ but $P(E)P(F) = .3(.45) = .135$

D.4 joint; marginal

D.5

| | Variable B | | |
Variable A	B_1	B_2	Total
A_1	.16	.04	.20
A_2	.56	.14	.70
A_3	.08	.02	.10
Total	.80	.20	1.00

D.6 Yes; if the variables are statistically dependent, at least some
(and perhaps all) of the joint probabilities would differ from
those determined in D.5. (Note: In the case of statistical
dependence, it is not possible to obtain the appropriate joint
probabilities from knowledge of the associated marginal
probabilities. They must be assessed directly from the underlying
problem setting or, if relevant conditional probabilities are
known, by the multiplication theorem.)

Problem 4

a.   i.   No such pairs of events are statistically independent
            since, for all Ai and Bj, $P(Ai \cap Bj) \neq P(Ai)P(Bj)$. For

ii.   Since not all joint probabilities are equal to the
        product of the corresponding marginal probabilities
        (indeed, in this case none of them are), the two
        variables are statistically independent.

b. i.

| | Conditional on Preliminary Cost Estimate | | |
	Under $1 Million B_1	$1 Million or More B_2	Total
Project Culmination			
A_1 : Affiliated GC	.067	.650	.300
A_2 : Different GC	.133	.050	.100
A_3 : Not Constructed	.800	.300	.600
Total	1.000	1.000	1.000

ii.   Projects with preliminary cost estimates $1 million or
        more are much more likely to be constructed by the
        affiliated general contractor than are projects with
        preliminary cost estimates under $1 million. Conversely,
        projects with preliminary cost estimates under $1
        million are more likely not to be constructed or to be
        constructed by a different general contractor.

5

RANDOM VARIABLES

A. BASIC CONCEPTS AND DISCRETE RANDOM VARIABLES
(Text: Chapter 5, Sections 5.1 and 5.2)

Review of Basic Concepts

A.1 When a sample space is composed of basic outcomes that are numerical, the characteristic of interest X is called a(n) _____ _____.

A.2 Determine whether each of the following sample spaces is related to a discrete or continuous random variable.

a. Time required to complete a slalom run in a ski competition.

   _____

b. Number of fire calls received per day at a central station.

   _____

c. Velocity of an automobile at the bottom of a steep grade.

   _____

A.3 A shipment arrives with a sealed carton containing four fragile ornaments. If random variable X denotes the number of broken ornaments in the carton, then the sample space of X has the numerical outcomes X = _____, _____, _____, _____, _____.

A.4 There are three employees in a small machine shop. The probability that no employees are on sick leave on a particular day is .91. Similar probabilities for 1, 2 or 3 employees on sick leave are .06, 02, and .01, respectively. Thus, X, the number of employees on sick leave is a discrete random variable. For this situation, explain what P(X = 2) or, simply, P(2) denotes. _____

A.5 For the situation in A.4, determine each of the following.

   a.   P(X = 0) = _____ .

   b.   P(3) = _____ .

   c.   P(X ≤ 1) = _____ .

   d.   P(X ≤ 2) = _____ .

   e. $\sum\limits_{x}$ P(X=x) = _____ .

A.6 The values of the probability P(X≤x), when considered for all values of x which can be assumed by a discrete random variable X, is called the _____ probability distribution of X.

A.7 The probability that random variable X equals some particular x and random variable Y equals some y is called a(n) _____ probability, and is denoted by _____ .

A.8 Explain what the symbol P(X=0|Y=3) denotes. _____
_____

A.9 P(X=0|Y=3) can be computed as _____ divided by
_____ .

A.10 The probability distribution for random variable X given that random variable Y has the value 3 is called a(n) _____ probability distribution.

A.11 Two random variables are statistically independent if and only if P(X=x ∩ Y=y) = _____ for all x and y.

A.12 A crisis intervention center receives 0, 1 or 2 calls in any given hour, with respective probabilities .5, .3, and .2. Let X and Y denote, respectively, the number of calls received in two successive hours. It is known that X and Y are statistically independent. Thus P(X=0 ∩ Y=0) = _____ , P(X=0 ∩ Y=1) = _____ , and P(X=0 ∩ Y=2) = _____ .

A.13 For the situation in A.12, P(Y=0|X=0) = _____ , P(Y=1|X=0)= _____ , and P (Y=2|X=0) = _____ .

»Problem 1

a. Consider the performance of three machines during a one-week
    production period. The random variable of interest is X, the
    number of machines that fail during the week. Machines which
    fail require more than a week to repair so multiple failures
    of the same machine are impossible. The following probability
    distribution for X is appropriate. ⊙

x	P(X=x)
0	.70
1	.15
2	.10
3	.05
Total	1.00

Construct graphs of the probability distribution of X and the
cumulative probability distribution of X. √

b. The bivariate probability distribution for X, the number of
    machines that fail during the week, and Y, the number of new
    operators assigned to the machines that week, is as follows.

		Number of New Operators y			
		0	1	2	Total
Number of	0	.661	.036	.003	.70
Machine	1	.126	.020	.004	.15
Failures	2	.082	.013	.005	.10
x	3	.031	.011	.008	.05
	Total	___	___	___	1.00

i.   Obtain the marginal probability distribution for Y.

ii. Are the random variables X and Y statistically
     independent? ____ Explain. _____ √

iii. Obtain the conditional probability distribution for X
     given that one new operator is assigned to the machines.

iv.   Obtain the conditional probability distribution for X given that two new operators are assigned to the machines. √

B. EXPECTATION AND VARIANCE, AND STANDARDIZED RANDOM VARIABLES

(Text: Chapter 5, Sections 5.3 and 5.4)

Review of Basic Concepts

B.1 The expected value of random variable X is denoted by the symbol _____ .

B.2 The expected value $E(X)$ is often referred to as the _____ of the probability distribution of X.

B.3 If random variable X can assume the values 1 or 2 with probabilities .8 and .2, respectively, then $E(X)$ = _____ .

B.4 In general, if a random variable X can assume the values 1 through k, the expected value of X is computed with the formula $E(X)$ = _____ .

B.5 The variance of the probability distribution of X is denoted by _____ .

B.6 If random variable X has five outcomes, explain in words how $\sigma^2(X)$ is computed using (5.6) in your Text. _____

_____

B.7 For the situation in B.3, $\sigma^2(X)$ is computed by summing the expressions _____ and _____ . Hence, $\sigma^2(X)$ = _____ .

B.8 If random variable X is measured in units of "telephone calls," in what units is $\sigma^2(X)$ measured? _____ .

B.9 The positive square root of $\sigma^2(X)$ is called the _____ _____ of X and is denoted by _____ .

B.10 Suppose X is a random variable with E(X) = 6 and $c^2(X)$ = 16.
Thus, $\sigma(X)$ = _____ .

B.11 The standardized form of random variable X is equal to X minus
_____ divided by _____ .

B.12 Suppose X is a random variable with E(X) = 10.0 and $\sigma(X)$ = 2.0.
Let Y denote the standardized form of X. When X assumes the value
7.0, the corresponding standardized value is Y = _____ .
Similarly, if X assumes the value 14.0, Y = _____ .

B.13 For the situation in B.12, E(Y) = _____ , $\sigma^2(Y)$ = _____ ,
and $\sigma(Y)$ = _____ .

»**Problem 2**

Patrons at a short-term (two-hour maximum) airport parking lot are
charged for each quarter-hour or fraction thereof that a vehicle is
parked. (That is, parking times are rounded up to the nearest
quarter-hour.) The probability distribution for X, the number of
quarter-hour periods charged to a parked vehicle, is as follows.

x:	1	2	3	4	5	6	7	8
P(x):	.05	.15	.22	.22	.17	.10	.06	.03

a.   Compute and interpret the meaning of E(X) for this situation.

_____

_____ √

b.   i.   Compute $\sigma^2(X)$, the variance of the probability
distribution of X. _____

ii.   Compute $\sigma(X)$, the standard deviation of the probability
distribution of X. _____ √

c.   i.   Obtain the probability distribution of Y, the
standardized form of the random variable X.

ii.   Obtain P (Y < -1.5) + P (Y > 1.5) from this probability
distribution.

C. CONTINUOUS RANDOM VARIABLES
(Text: Chapter 5, Section 5.5)

Review of Basic Concepts

C.1 For continuous random variables, probability is associated with
_____ on a continuum rather than with points on a
continuum.

C.2 The probability that X will take on a value in any given interval
is given as the corresponding area under the curve of the
_____ _____ _____ for X.

C.3 Suppose the probability density function for X comprises a
rectangle of height .25 in the interval $0 \leq x \leq 4$. Write the
probability density function for X symbolically, $f(x) =$
_____ .

C.4 Recall that the area of a rectangle is equal to its width
multiplied by its height. Thus, for the situational in E.3, the
total area under the curve of $f(x)$ is _____. Further,
$P(0 \leq X \leq 2) =$ _____ and $P(3 \leq X \leq 4) =$ _____ .

C.5 For a continuous random variable, F(x) denotes the _____
_____ _____ of X.

C.6 If X is a continuous random variable, interpret the result $F(3.0) =$
.75. _____ √

»Problem 3

The probability density function and cumulative probability
function for x, the time (in minutes) needed to swim the required
distance in the final exam for an intermediate swimming class, are as
follows.

$$f(x) = \begin{cases} .5 & 1 \leq x \leq 3 \\ 0 & \text{elsewhere} \end{cases} \qquad F(x) = \begin{cases} 0 & x < 1 \\ .5(x - 1) & 1 \leq x \leq 3 \\ 1 & x > 3 \end{cases}$$

a. Construct graphs of the probability density function and the
cumulative probability function for X. √

b. Determine each of the following from your graphs in part (a.).
    (Note: The area of a rectangle is equal to its width
    multiplied by its height.)

    i. $f(2)$ = _____ ;
          $f(2.5)$ = _____

    ii. $F(1.5)$ = _____ ;
          $F(2.5)$ = _____ ;
          $F(3.0)$ = _____ √

c. Verify the preceding formula for $F(x)$. _____

_____

Answers to Chapter Reviews and Problems

A.1 random variable

A.2 a. continuous

     b. discrete

     c. continuous

A.3 x = 0, 1, 2, 3, 4

A.4 $P(X = 2)$ or $P(2)$ denotes the probability that two employees are on
     sick leave.

A.5 a. $P(X = 0)$ = .91;

     b. $P(3)$ = .01

     c. $P(X \le 1)$ = .91 + .06 = .97

     d. $P(X \le 2)$ = .91 + .06 + .02 = .99

     e. $\sum_x P(X = x)$ = $P(X = 0) + P(X = 1) + P(X = 2) + P(X = 3)$ = 1.0

A.6 cumulative

A.7 joint; $P(X = x \cap Y = y)$

A.8 P $(X = 0 | Y = 3)$ denotes the conditional probability that X assumes the value 0 given that Y assumes the value 3.

A.9 P $(X = 0 \cap Y = 3)$; $P(Y = 3)$

A.10 conditional

A.11 $P(X = x)P(Y = y)$

A.12 $P(X = 0 \cap Y = 0) = P(X = 0)P(Y = 0) = .5(.5) = .25$;
$P(X = 0 \cap Y = 1) = .5(.3) = .15$;
$P(X = 0 \cap Y = 2) = .5(.2) = .10$

A.13 $P(Y = 0 | X = 0) = .25/.50 = .50$;
$P(Y = 1 | X = 0) = .15/.50 = .30$;
$P(Y = 2 | X = 0) = .10/.50 = .20$

(Note: Because X and Y are statistically independent here, $P(Y = y | X = x) = P(Y = y)$. For example, $P(Y = 0 | X = 0) = P(Y = 0) = .5$.

Problem 1

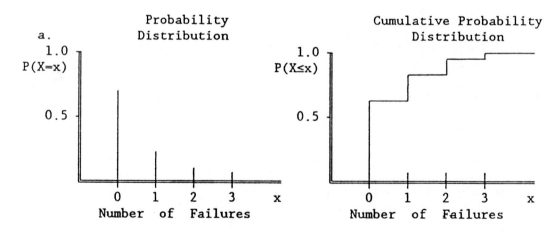

Figure 5.1

b. i. $P(Y = 0) = .90$; $P(Y = 1) = .08$; $P(Y = 2) = .02$.

ii.   No. For example, P(X = 0 ∩ Y =0) = .661
      whereas P(X = 0)P(Y = 0) = .70(.90) = .630.

iii.

Number of Machine Failures x	Conditional Probability P(X = x|Y = 1)
0	.036/.08 = .4500
1	.020/.08 = .2500
2	.013/.08 = .1625
3	.011/.08 = .1375
Total	1.0000

iv.

x	P(X =x|Y = 2)
0	.15
1	.20
2	.25
3	.40
Total	1.00

B.1 E(X)

B.2 mean

B.3 E(X) = 1(.8) + 2(.2) = 1.2

B.4 $E(X) = \sum_x xP(x)$

B.5 $\sigma^2(X)$

B.6 E(X) is computed first. The deviation of each outcome from E(X) is computed next. Each of these five deviations ·is then squared. Next, each squared deviation is weighted by (multiplied by) the probability of that outcome occurring. Finally, the five weighted squared deviations are summed to yield $\sigma^2(X)$.

B.7 $(1 - 1.2)^2(.8)$; $(2 - 1.2)^2(.2)$; $\sigma^2(X) = .16$.

B.8 telephone calls squared

B.9 standard deviation; $\sigma(X)$

B.10 $\sigma(X) = +\sqrt{16} = 4$

B.11 $E(X)$; $\sigma(X)$

B.12 $Y = \dfrac{7.0 - 10.0}{2.0} = -1.5$; $Y = \dfrac{14.0 - 10.0}{2.0} = 2.0$

B.13 $E(Y) = 0$; $= \sigma^2(Y) = 1$; $\sigma(Y) = 1$

Problem 2

a. $E(X) = 1(.05) + 2(.15) + 3(.22) + 4(.22) + 5(.17) +6(.10)$ $+7(.06) + 8(.03) = 4.0$; $E(X)$ can be interpreted as the mean number of quarter-hour periods for which vehicles are charged, over many independent vehicle parking times.

b. i. $\sigma^2(X) = (1 - 4)^2(.05) + (2 - 4)^2(.15) + (3 - 4)^2(.22) +(4 - 4)^2(.22) + (5 - 4)^2(.17) + (6 - 4)^2(.10) + (7 - 4)^2(.06) + (8 - 4)^2(.03) = 2.86$.

    ii. $\sigma(X) = \sqrt{2.86} = 1.691$

c. i. $E(X) = 4$; $\sigma(X) = 1.691$; Thus, from (5.9) in your Text we obtain the standardized form $Y = (X - 4)/1.691$. Hence, the probability distribution for Y appears as follows:

y:	-1.77	-1.18	-.59	0	.59	1.18	1.77	2.37
P(y):	.05	.15	.22	.22	.17	.10	.06	.03

    ii. $P(Y < -1.5) + P(Y > 1.5) = .05 + .06 + .03 = .14$

C.1 intervals

C.2 probability density function

C.3 $f(x) = \begin{cases} .25 & 0 \le x \le 4 \\ 0 & \text{elsewhere} \end{cases}$

C.4 $4(.25) = 1.0$; $2(.25) = .50$; $1(.25) = .25$

C.5   cumulative probability function

C.6 The probability that X will assume the value 3.0 or less is equal
     to .75.

Problem 3

a.

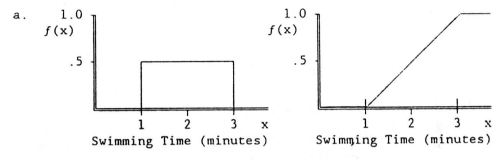

Figure 5.2

b.   i. $f(2) = .5; f(2.5) = .5$

    ii. $F(1.5) = .25; F(2.5) = .75;: F(3.0) = 1.0$

c.   Using geometry, the area under the probability density
     function to the left of any value x in the interval $1 \leq x \leq 3$
     is $.5(x - 1)$. Hence, $F(x) = .5(x - 1)$ for $1 \leq x \leq 3$.

6

ADDITIONAL TOPICS IN PROBABILITY

A. BAYES' THEOREM
(Text: Chapter 6, Section 6.1)

Review of Basic Concepts

A.1 Bayes Rule is an application of the _____ _____.

A.2 Probabilities calculated by Bayes Rule are called _____ probabilities, since they are bases on _____ _____.

»Problem 1

Among all schools which offer programs in a particular field of study, 30 percent have been accredited nationally. Following are the conditional probability distributions for highest degree offered by a school in this field of study given accreditation status.

Highest Degree Offered	Probability Conditioned Upon Accredited	Probability Conditioned Upon Not Accredited
Bachelor of Arts	.15	.85
Master of Arts	.25	.10
Doctor of Philosophy	.60	.05
	1.0	1.0

a. Use Bayes rule to calculate the probability that a randomly selected school is accredited given it offers just a Bachelor of Arts degree in this field. _____

b. Use Bayes rule to calculate the probability that a randomly selected school is not accredited given it offers a Doctor of Philosophy degree in this field. _____

B.   SUMS AND DIFFERENCES OF TWO INDEPENDENT RANDOM VARIABLES
(Text: Chapter 6, Section 6.2)

Review of Basic Concepts

B.1 The expected value of the sum of two independent random variables, $E(X + Y) =$ _____ and the expected value of the difference between two independent random variables, $E(X - Y) =$ _____ .

B.2 If X and Y are independent, $\sigma^2(X + Y) =$ _____ and $\sigma^2(X - Y) =$ _____ .

B.3 If X, Y and Z are three independent random variables, $E(X + Y + Z) =$ _____ and $\sigma^2(X + Y + Z) =$ _____ .

B.4 A firm has four identical warehouses, each with the same probability distribution of loss by fire. The mean and standard deviation of this probability distribution (in $ thousands) are 50 and 30, respectively. If fire losses in the four warehouses are statistically independent, find the mean, variance, and standard deviation of T, the total fire loss of the four warehouses.

a. $E(T) =$ _____ ;

b. $\sigma^2(T) =$ _____ ;

c. $\sigma(T) =$ _____ . ✓

»Problem 2

Speeding violations in a locality result in fines of either $35, $45, $55, or $65, depending upon the speed in excess of the posted limit. The probability distribution for X, the fine for a violation, is as follows.

x:	35	45	55	65
P(x):	.50	.10	.30	.10

The mean and variance of X are $E(X) = 45.0$ and $\sigma^2(E) = 120.0$. Let X and Y denote the fines for the next two speeding violations. Assume that X and Y are statistically independent.

a. Obtain the bivariate probability distribution for X and Y. √

b. Using your results in part (a.) , obtain the probability distribution for the random variable $Z = X + Y$, the total amount of fines for the two speeding violations. √

c. i. Using the probability distribution in part (b.), compute $E(Z)$ and $\sigma^2(Z)$.

   $E(Z)$ = _____ ; $\sigma^2(Z)$ = _____ .

   ii. Obtain $E(Z)$ and $\sigma^2(Z)$ by using (6.7) in your Text.

   $E(Z)$ = _____ ; $\sigma^2(Z)$ = _____ . √

C. FUNCTIONS OF RANDOM VARIABLES
(Text: Chapter 6, Section 6.2)

Review of Basic Concepts

C.1 If X is a random variable, then the random variable $W = a + bX$ is called a(n) _____ _____ of X.

C.2 Suppose X is a random variable and $W = 2 + 6X$. If $P(X = 2) = .4$, then $P(W = 14)$ = _____ . Further, if $E(X) = 3.5$, then $E(W)$ = _____ .

C.3 The expected value $E(a)$ of any constant a equals _____ .

C.4 Suppose X is a random variable with $E(X) = 6$ and $\sigma^2(X) = 16$. If $W = 3 + 5X$, then, $E(W)$ = _____ , $\sigma^2(W)$ = _____ , and $\sigma(W)$ = _____ .

C.5 Given a random variable X with $E(X) = 3.0$ and $\sigma^2(X) = 1.5$, determine $\sigma^2(W)$ for each of the following linear functions of X.

   a. $W = 7 + 10X$: _____

   b. $W = 7 - 10X$: _____

   c. $W = 10X$: _____

d.   W = 7 + X: _____ √

»**Problem 3**

Refer to the situation in Chapter 5, Problem 2, which introduced the random variable X, the number of quarter-hour periods for which a vehicle is charged at a short-term airport parking lot. Recall that E(X) = 4.0 and $\sigma^2(X)$ = 2.86.

a. The current parking rate is $.30 for each quarter-hour or fraction thereof.

i.   Identify the linear function relating X to W, the dollar amount received for a vehicle parked in this lot. _____√

ii.   Obtain the probability distribution for W. √

iii. Compute E(W) and $\sigma^2(W)$ directly from the probability distribution in part (ii.).

E(W) = _____; $\sigma^2(W)$ = _____.

iv. Verify E(W) and $\sigma^2(W)$ by using (6.5b) and (6.6b) in your Text.

E(W) = _____; $\sigma^2(W)$ = _____.

v.   Obtain $\sigma(W)$. _____ √

b. A new parking rate of $.55 for the first quarter-hour and $.25 for each additional quarter-hour or fraction thereof has been proposed. For this situation, W, the dollar amount received, is related to X, parking time charged to a vehicle, by the linear function W = .30 + .25X over the relevant range of values for X. What impact would the new rate have on expected receipts and variance of receipts per vehicle parked, assuming the new rate would not affect the probability distribution of X?

_____ √

D.   COVARIANCE AND CORRELATION
(Chapter 6, Section 6.3)

The following data on the number of sales personnel (variable X) and daily unit of sales (variable Y) were observed at a dress shop. The outcome probabilities are assumed equal for each of these eight randomly selected days, i.e., P(x,y) = .125 for each day.

Y:	13	10	9	12	11	15	14	12
X:	6	3	5	7	8	11	9	7

D.1 What was the average daily sales? _____

D.2 What was the average size of the sales staff? _____

D.3 Estimate the amount by which daily sales varies on average by calculating the standard deviation of sales. _____

D.4 Estimate the amount by which sales staff varies on average by calculating the standard deviation of staff size. _____

D.5 Calculate the covariation between sales and personnel. _____

D.6 Calculate the correlation coefficient between sales and personnel. _____. Interpret this number._____

_____

E.   CALCULATIONS FOR CONTINUOUS RANDOM VARIABLES
(Chapter 6, Section 6.4)

E.1 For the situation in Chapter 5, Problem 3, verify the formula for F(x) using calculus _____

_____

E.2 For the same situation, determine the expected value and variance X.

E(X)=_____; $\sigma^2(X)$= _____ √

Answers to Chapter Reviews and Problems

A.1 Multiplication rule

A.2 posterior; additional information

Problem 1

A_i	$P(A_i)$	B_j	$P(B_j|A_i)$	$P(A_i)P(B_j|A_i)$
		B.A.	.15	.045
Accredited	.30	M.A.	.25	.075
		Ph.D.	.60	.180
Not		B.A.	.15	.045
Accredited	.70	M.A.	.25	.075
		Ph.D.	.60	.180

Tree Diagram of Probabilities
Figure 6.1

a.
$$P(A_1|B_1) = \frac{P(A_1)\ P(B_1|A_1)}{P(A_1)P(B_1|A_1) + P(A_2)P(B_1|A_2)} = \frac{.045}{.045 + .595} = .0703125$$

b.
$$P(A_2|B_3) = \frac{P(A_2)\ P(B_3|A_2)}{P(A_1)P(B_3|A_1) + P(A_2)P(B_3|A_2)} = \frac{.035}{.180 + .035} = .1627906$$

B.1 $E(X + Y) = E(X) + E(Y)$; $E(X - Y) = E(X) - E(Y)$

B.2 $\sigma^2(X+Y) = \sigma^2(X) + \sigma^2(Y)$; $\sigma^2(X-Y) = \sigma^2(X) + \sigma^2(Y)$

B.3 $E(X+Y+Z) = E(X) + E(Y) + E(Z)$; $\sigma^2(X+Y+Z) = \sigma^2(X) + \sigma^2(Y) + \sigma^2(Z)$

B.4 a. $E(T) = 50 + 50 + 50 + 50 = 200$

b. $\sigma^2(T) = (30)^2 + (30)^2 + (30)^2 + (30)^2 = 3,600$

c. $\sigma(T) = \sqrt{3600} = 60$

Problem 2

a.

		y			
	35	45	55	65	Total
35	.25	.05	.15	.05	.50
45	.05	.01	.03	.01	.10
x 55	.15	.03	.09	.03	.30
65	.05	.01	.03	.01	.10
Total	.50	.10	.30	.10	1.00

b. If $X = 35$ and $Y = 35$, then $Z = 70$. Thus, $P(Z = 70) = P(X = 35 \cap Y = 35) = .25$. Similarly, if $X = 35$ and $Y = 45$, or if $X = 45$ and $Y = 35$, then $Z = 80$. Thus, $P(Z = 80) = P(X = 35 \cap Y = 45) + P(X = 45 \cap Y = 35) = .05 + .05 = .10$. Continuing in this fashion, we obtain the following probability distribution for Z.

z:	70	80	90	100	110	120	130
P(z):	.25	.10	.31	.16	.11	.06	.01

c. i. $E(Z) = 70(.25) + 80(.10) = \ldots + 130(.01) = 90.0$; $\sigma^2(Z) = (70-90)^2(.25) + (80-90)^2(.10) + \ldots + (130-90)^2(.01) = 240.0$

ii. $E(X) = E(Y) = 45.0$; $\sigma^2(X) = \sigma^2(Y) = 120.0$; Thus, $E(Z) = 45.0 + 45.0 = 90.0$ and $\sigma^2(Z) = 120.0 + 120.0 = 240.0$.

C.1 linear function

C.2 .4; $E(W) = 2 + 6(3.5) = 23.0$

C.3 a

C.4 $E(W) = 3 + 5(6.0) = 33.0$; $\sigma^2(W) = (5)^2(16) = 400$; $\sigma(W) = 20$

C.5 a. $\sigma^2(W) = (10)^2(1.5) = 150$

   b. $\sigma^2(W) = (-10)^2(1.5) = 150$

   c. $\sigma^2(W) = (10)^2(1.5) = 150$

d. $\sigma^2(W) = (1)^2(1.5) = 15$

Problem 3

a. i. $W = .30X$

   ii.

w:	.3	.6	.9	1.2	1.5	1.8	2.1	2.4
P(w):	.05	.15	.22	.22	.17	.10	.06	.03

   iii. $E(W) = .3(.05) + .6(.15) + 9(.22) + 1.2(.22) + 1.5(.17)$
   $+ 1.8(10) + 2.1(0.6) + 2.4(.03) = \$1.20$

   $\sigma^2(W) = (.3-1.2)^2(.05) + (.6-1.2)^2(.15) + (.9-1.2)^2(.22)$
   $+(1.2-1.2)^2(.22) + (1.5-1.2)^2(.17) + (1.8-1.2)^2(.10)$
   $+ (2.1-1.2)^2(.06) + (2.4 - 1.2)^2(.03) = .2574$

   iv. $E(W) = .3(4.0) = \$1.20$
   $\sigma^2(W) = (.30)^2(2.86) = .2574$

   v. $\sigma(W) = \sqrt{.2574} = .5073$

b. Under the proposed new rate, $E(W) = .30 + .25(4) = \$1.30$ and $\sigma^2(W) = (.25)^2(2.86) = .1788$. Therefore; expected receipts per vehicle would increase by \$.10, from \$1.20 to \$1.30, while the variance would decrease slightly, from .2574 to .1788.

D.1 $\bar{X} = \Sigma X/n = 56/8 = 7$ staff

D.2 $\bar{Y} = \Sigma Y/n = 96/n = 12$ dresses

D.3 $s^2(X) = \Sigma(X-\bar{X})^2/(n-1) = 42/7 = 6$, so $s(X) = \sqrt{6} =$

D.4 $s^2(Y) = \Sigma(Y-\bar{Y})^2/(n-1) = 28/7 = 4$, so $s(Y) = \sqrt{4} = 2$

D.5 $\Sigma(Y-\bar{Y})^2\Sigma(X-\bar{X}) = 28$, so covariance is $s(X,Y) = 4$

D.6 $r = s(X,Y)/s(X)s(Y) = 4/2\sqrt{6} = .6667$; about two-thirds of the variation in Y is associated with variation in X.

E.1 Using geometry, the area under the probability density function to the left of any value x in the interval $1 \leq x \leq 3$ is $.5(x - 1)$. Hence, $F(x) = .5(x - 1)$ for $1 \leq x \leq 3$. Note: Using calculus, we obtain

$$F(x) = \int_1^x .5du = .5u \Big|_1^x = .5x - .5(x - 1) \text{ for } 1 \leq x \leq 3$$

E.2 $E(x) = \int_1^3 xf(x) \, dx = \int_1^3 x \, (.5)dx = .25x^2 \Big|_1^3 = 2.25 - .25 = 2.00;$

$$\sigma^2(x) = \int_1^3 [x - E(X)]^2 \, f(x) \, dx = \int_1^3 (x - 2)^2 \, (.5) \, dx$$

$$= \int_1^3 (.5x^2 - 2x + 2) \, dx$$

$$= (x3/6) - x^2 = 2x \Big|_1^3 = (9/6) - (7/6) = 2/6 = .333$$

7

COMMON PROBABILITY DISTRIBUTIONS DISCRETE RANDOM VARIABLES

A. DISCRETE UNIFORM PROBABILITY DISTRIBUTIONS
(Text: Chapter 7, Section 7.1)

Review of Basic Concepts

A.1 To obtain a discrete uniform probability distribution from the family of all such distribution, we must specify two parameters; a and s. Explain in words what each of these parameters denotes.
_____

A.2 Given a discrete uniform probability distribution with parameters a = 1 and s = 12, P(X=6) = P(6) = _____.

A.3 For the situation in A.2, E(X) _____ and $\sigma^2(X)=$ _____.

Problem 1

Ages (to the nearest year) of students in an elementary school, X, follow a discrete uniform probability distribution with possible values from 6 to 13, inclusive.

    a. Determine each of the following for this distribution.

        i. a = _____; s = _____

       ii. P(X = x) = P(x) = _____

     iii. E(X)= _____; $\sigma^2(X)=$ _____

    b. Determine each of the following probabilities.

        i. Probability that a student selected at random is less than 9 years old. _____

       ii. Probability that a student selected at random is 10 years old or more. _____

B. BINOMIAL PROBABILITY DISTRIBUTIONS
(Text: Chapter 7, Section 7.2)

Review of Basic Concepts

B.1 A random trial that has two basic outcomes of a qualitative nature is called a (n) _____ random trial.

B.2 Indicate whether or not each of the following represents a Bernoulli random variable.

    a.   Time between two successive arrivals at an airline ticket counter, measured to the nearest minute. _____

    b.   Whether or not an employee commutes to work in a car pool. _____

    c.   Number (actual count) of faulty cigarette filters in a sample of twenty filters. _____

    d.   A lab test which may be positive or negative. _____

B.3 One condition, or postulate, that makes a sequence of Bernoulli trials a Bernoulli process is that the trials in the sequence are and $P(Xi = 0)$ are the same for all trials. This second condition is called _____ .

B.4 If the random variables in any sequence are statistically independent of one another and if each has the same probability distribution, the random variables are said to be independent and _____ .

B.5 In a statistical study, a Bernoulli random variable is defined as $X_i = 1$ if the ith defendant accused of shoplifting is convicted and $X_i = 0$ if acquitted. Eighty percent of all such defendants are convicted. A Bernoulli process of 10 defendants is being studied. Define X, the relevant binomial random variable of the study.

_____

B.6 For the situation in B.5, write the expression which must be solved to compute the probability that six of the ten defendants will be convicted. _____ .

B.7 Solve each of the following binomial coefficients.

a. $\binom{5}{4}$ = _____   c. $\binom{6}{2}$ = _____
b. $\binom{4}{2}$ = _____   d. $\binom{5}{5}$ = _____

B.8 For the situation in B.5 and B.6, P(6) = _____

B.9 To obtain a binomial probability distribution from the family of all such distributions, we must assign values for _____ and _____. These are called the _____ of the binomial probability distribution.

B.10 In which direction is the binomial probability distribution skewed, if at all, for the following p values.

a. p = .15 _____

b. p = .80 _____

c. p = .50 _____

B.11 Given the parameters n and p of a binomial distribution, the mean of the distribution, E(X), equals _____ and the variance, $\sigma^2(X)$, equals _____. Thus, for the situation in B.5, E(X) = _____ and $\sigma^2(X)$ = _____.

B.12 Use Table C-5 in your Text to determine each of the following binomial probabilities.

a. P(X = 6) if n - 10, P - .35. _____

b. P(X = 4) if n = 10, p = .65. _____

c. P(X ≤ 3) if n - 5, p = .25. _____

d. P(X ≥ 3) if n - 4, p - .93. _____

B.13 Let random variable X be defined as in B.5. Let Y be the number who are convicted among a further 5 defendants from the same Bernoulli process. Therefore, X and Y are independent binomial random variables each with p - .8. Thus, the sum X+Y is a (n) _____ random variable with parameters p = _____ and n = _____.✓

»Problem 2

In a marketing experiment, a man is shown a pair of belts, a pair of shirts, and a pair of suits, and for each pair is asked to identify the more expensive item of the two. Assume that the man has a probability of .5 of identifying the more expensive item correctly (i.e., he is guessing) and that each identification is independent of the others.

a. Use Table C-5 in your Text to obtain the probability distribution for X, the number of times the more expensive item is correctly identified.

b. Verify $P(X=1)$ by using (7.6) in your Text. _____ √

c. Determine the mean and variance of the probability distribution of X by using (7.72) and (7.76) in your Text. $E(X)=$ _____ $\sigma^2(X)=$ _____ √

d. Suppose this experiment is expanded by adding an additional pair of items.

   i. Use Table C-5 in your Text to determine the probability of exactly one correct identification out of the four pairs. _____ √

   ii. Why is this probability smaller than the probability $P(X = 1)$ in part (a.)? _____ _____ √

C. POISSON PROBABILITY DISTRIBUTIONS
(Text: Chapter 7, Section 7.3)

Review of Basic Concepts

C.1 To obtain a Poisson probability distribution from the family of all such distributions, we must assign a value for _____, the parameter of the distribution.

C.2 Is it true that the parameter of a Poisson probability distribution must be a positive integer? _____ Explain. _____

C.3 The parameter of a Poisson probability distribution is 4.0. The mean of this distribution is _____ , the variance is _____ and the standard deviation is _____ .

C.4 All Poisson probability distributions are skewed to the _____ .

C.5 Given a Poisson probability distribution with $\lambda = 6.0$, write the expression which must be solved to compute $P(X = 3)$. _____

C.6 In Section 7.3 of your Text, the postulates of a Poisson process are illustrated for a process generating occurrences over time. As an example of a process not related to time, consider a process generating typing mistakes on a page. In this instance, instead of subdividing time into small non-overlapping intervals of size Δt, we could subdivide a page into small non-overlapping intervals of typed characters. In your own words, illustrate the four postulates which would make this a Poisson process.

a. Postulate 1: _____

b. Postulate 2: _____

c. Postulate 3: _____

d. Postulate 4: _____

C.7 Use Table C-6 in your Text to determine each of the following Poisson probabilities.

a. $P(X = 7)$ if $\lambda = 5.5$. _____

b. $P(X = 7)$ if $\lambda = 17.0$. _____

c. $P(X \leq 3)$ if $\lambda = 4.0$. _____

d. $P(X \geq 5)$ if $\lambda = 2.0$. _____

C.8 The sum of two independent Poisson random variables with parameters $\lambda_1 = 2.0$ and $\lambda_2 = 3.0$ is a _____ random variable with parameter $\lambda = $ _____ . ✓

»Problem 3

The number of patients who arrive each hour at the emergency room of a metropolitan hospital, X, is a Poisson random variable with $\lambda = .5$.

a. Use Table C-6 in your Text to obtain the probability distribution of X.

b. Verify P(2) in the probability distribution of X by using (7.8) in your Text. (Note: exp(-.5) = .6065). _____ √

c. Obtain the mean and variance of X by using (7.92) and (7.96) in your text.

   E(X) = _____ ; $\sigma^2(X)$ = _____ √

d. Additional staff must be called into the emergency room whenever more than two patients arrive in one hour. Obtain the probability that more than two patients will arrive in one hour. _____ √

Answers to Chapter Reviews and Problems

A.1 The parameters a and s denote the smallest outcome and the number of distinct outcomes, respectively, for a discrete uniform random variable.

A.2 1/12 = .083

A.3 $E(X) = 1 + \dfrac{12-1}{2} = 6.5$; $\sigma^2(X) = \dfrac{12^2-1}{12} = 11.917$

Problem 1

i.   a = 6; s = 8

ii. P(x) = 1/8 = .125 for x = 6, 7,...,13

iii. $E(X) = 6 + \dfrac{8-1}{2} = 9.5$; $\sigma^2(X) = \dfrac{8^2-1}{12} = 5.25$

b.i. $P(X < 9) = .375$

 ii. $P(X \geq 10) = .50$

B.1 Bernoulli

B.2 a. No;   b. Yes;   c. No.;   d. Yes

B.3 stationarity

B.4 identically distributed

B.5 The binomial random variable of interest, X, is the number of defendants who are convicted in the group of ten being studied; $X = X_1 + X_2 + \ldots + X_{10}$.

B.6 $P(X = 6) = P(6) = \binom{10}{6} (.8)^6 (1-8)^{10-6}$

B.7 a. $\dfrac{5!}{4!1!} = 5$;   b. $\dfrac{4!}{2!2!} = 6$;   c. $\dfrac{6!}{2!4!} = 15$;   d. $\dfrac{5!}{5!10!} = 1$

B.8 $P(6) = 210(.8)^6(.2)^4 = .0881$

B.9 n; p; parameters

B.10 a. right;   b. left;   c. unskewed or symmetrical

B.11 $E\{X\} = np$; $\sigma^2\{X\} = np(1 - p)$; $E\{X\} = 10(.8) = 8.0$; $\sigma^2\{X\} = 10(.8)(1 - .8) = 1.6$

B.12 a. .0689;   b. .0689;   c. .9844;   d. .9733

B.13 binomial; $p = .8$; $n = 10 + 5 = 15$

Problem 2

a. $n = 3$; $p = .5$; Thus, the probability distribution of X is as follows:

x:	0	1	2	3
P(x):	.125	.375	.375	.125

b. $P(X = 1) = P(1) = \binom{3}{1}(.5)^1(.5)^{3-1} = .375$

c. $E\{X\} = 3(.5) = 1.5$; $\sigma^2(X) = 3(.5)(1 - .5) = .75$

d. i. $n = 4$; $p = .5$; Thus, $P(1) = .250$.

    ii. Although the number of sequences containing one correct identification has increased from 3 to 4 (by the binomial coefficient), the probability of obtaining any one of these sequences with $p = .5$ has decreased from $.5^1.5^{3-1} = .125$ to $.5^1.5^{4-1} = .0625$.

C.1 λ

C.2 No; the parameter must be a positive number, but need not be an integer.

C.3 $E\{X\} = 4.0$; $\sigma^2(X) = 4.0$; $\sigma(X) = 2.0$

C.4 right

C.5 $P(X=3) = P(3) = \dfrac{6^3 \exp(-6)}{3!}$

C.6 a. The numbers of mistakes in small non-overlapping intervals of typed characters are statistically independent.

    b. The number of mistakes in an interval has the same probability distribution for all such intervals.

    c. The probability of one mistake in an interval is approximately proportional to the size of the interval.

d. The probability of two or more mistakes in an interval is very small compared to the probability of one mistake in that interval.

C.7 a. .1234;   b. .0034;   c. .4335;   d. .0526

C.8 Poisson; λ = 5.0

Problem 3

x:	0	1	2	3	4	5
a. P(x) :	.6065	.3033	.0758	.0126	.0016	.0002

b. $P(2) = \dfrac{.5^2 \exp(-5)}{2!} = .25(.6065)/2 = .0758$

c. $E\{X\} = \lambda = .5;\ \sigma^2\{X\} = \lambda = .5$

d. $P(x > 2) = .0126 + .0016 + .0002 = .0144$

8

COMMON PROBABILITY FUNCTIONS: CONTINUOUS RANDOM VARIABLES

A. CONTINUOUS UNIFORM (RECTANGULAR) PROBABILITY DISTRIBUTIONS

(Text: Chapter 8, section 8.1)

Review of Basic Concepts

A.1 To obtain a continuous uniform probability distribution from the family of all such distributions, we must specify two parameters. Identify these parameters. _____

_____

A.2 Given a continuous uniform probability distribution with a = 10 and b = 14, the appropriate probability density function is $f(x)$ = _____. Describe the shape of this distribution as seen in a graph of the probability density function $f(x)$.

_____

_____

A.3 For the situation in A.2, E(X) = _____ and $\sigma^2(X)$ = _____.

A.4 For the situation in A.2, the appropriate cumulative probability function is F(x) = _____. Hence, P(X≤11.5) = F(11.5) = _____.√

»Problem 1

As part of an exercise in spatial perception, preschool children are given a circular device which can be expanded from a minimum diameter of 10 centimeters to a maximum diameter of 20 centimeters. Using this device, a child is asked to estimate the size of a ball situated across the room. Let the random variable X denote the diameter of a child's first estimate. From past experience it is known that, for any child, X follows a continuous uniform probability distribution with parameters a = 10 and b = 20.

a. Obtain the probability density function for X. _____ √

b. i. Construct a graph of the probability distribution of X.

   ii. Determine the mean and variance for this distribution.

   E(X)= _____ ; $\sigma^2(X)=$ _____ √

c. Find each of the following probabilities for a randomly selected child who is making a first estimate.

   i. Probability that the diameter of his estimate is 13.5 centimeters or less. _____.√

   ii. Probability that the diameter of his estimate is greater than 16.0 centimeters. _____

   iii. Probability that the diameter of his estimate is between 13.5 and 16.0 centimeters. _____ √

B. NORMAL PROBABILITY DISTRIBUTIONS
(Text: Chapter 8, Section 8.2)

Review of Basic Concepts

B.1 To obtain a normal probability distribution from the family of normal probability distributions, two parameters must be specified. These are the mean of the distribution, denoted by _____, and the standard deviation of the distribution, denoted by _____.

B.2 Does the skewness of a normal probability distribution depend on the values of its parameters? _____ Explain. _____
_____

B.3 Does the spread or variability of a normal probability distribution depend on the values of its parameters? _____ Explain. _____
_____

B.4 Does the center (i.e., position on the x-axis) of a normal probability distribution depend on the values of its parameters? _____ Explain. _____
_____

B.5 N(5, 7) denotes a(n) _____ probability distribution with
_____ of 5 and _____ of 7.

B.6 In compact notation, N(μ, σ^2), the standard normal probability
distribution is denoted by _____. The standard
normal variable is denoted by _____.

B.7 In Table C-1 in your Text, the cell entry corresponding to the
value Z = 1.17 is .8790. Hence P(Z_____) = .8790.
Similarly, the 87.90th percentile of the standard normal distribution, which is denoted by _____, is equal to _____.

B.8 Since all normal probability distributions are continuous,
P(Z $\leq$ z) = P(Z < z). Hence, P(Z $\leq$ .47) = P(Z < .47) = _____.

B.9 Using Table C-1 in your Text, P(Z $\leq$ 1.96) = _____ and
P(Z $\leq$ 1.60) = _____. Hence, P(1.60 $\leq$ Z $\leq$ 1.96) = P(Z $\leq$ 1.96) -
P(Z $\leq$ 1.60) = _____.

B.10 By the complementation theorem, P(Z $\geq$ 1.46) = 1 - P(_____) =
_____.

B.11 Since all normal probability distributions are symmetrical,
P(Z $\geq$ -z) = P(Z $\leq$ z). Hence, P(Z $\geq$ -1.92) = P(Z $\leq$_____) =
_____. By the same reasoning, P(-1.92 $\leq$ Z $\leq$ -1.40) = P(1.40 $\leq$
Z $\leq$_____) = _____.

B.12 Using the symmetry of the normal probability distribution and the
complementation theorem, P(Z $\leq$ -.64) = P(Z $\geq$ _____) =
1 - P(_____) = _____.

B.13 Use Table C-1 in your Text to determine each of the following
probabilities for Z, the standard normal variable.

a. P(1.18 $\leq$ Z $\leq$ 1.74) = _____

b. P(Z $\leq$ -2.32) = _____

c. P(-2.32 $\leq$ Z $\leq$ 1.18) = _____

d. P(-2.32 $\leq$ Z $\leq$ -.36) = _____

B.14 Use Table C-1 in your Text to determine each of the following percentiles of the standard normal distribution.

a. 83.4th percentile _____

b. 50th percentile _____

c. 2.5th percentile _____

d. 98th percentile _____

B.15 If X is a normal random variable with $\mu=10.0$ and $\sigma^2=2.0$, the value for z corresponding to x = 13.5 is _____. Hence, P(X ≤ 13.5) = P(Z_____) = _____.

B.16 For the situation in B.15, P(9.0 ≤ X ≤ 13.5) = P(_____Z_____) = _____.

B.17 For the situation in B.15, determine each of the following probabilities.

a. P(11.0 ≤ X ≤ 13.5) = _____

b. P(X > 13.6) = _____

c. P(X ≤ 8.2) = _____

d. P(6.0 ≤ X ≤ 14.0) = _____

e. P(X > 7.5) = _____

f. P(6.8 ≤ X ≤ 7.5) = _____

B.18 Suppose X and Y are independent random variables such that X is N(30,25) and Y is N(40,100). The sum X + Y is a(n) _____ random variable with mean _____ and variance _____ √

»Problem 2

The revenue a charity will realize in the coming year, X, is a normal random variable with μ = \$525 million and σ^2 = \$20 million.

a. It is desired to obtain the probability that revenue will exceed
   $500 million.

   i. Construct a graph similar to Figure 8.6 in your Text showing
      the probability distribution of X, the desired probability,
      and the correspondence between the probability distributions
      of X and Z.√

   ii. Obtain P(X > 500)._____√

b. Determine each of the following probabilities for X. (Note: It
   may be helpful to draw a rough sketch showing the desired
   probability for each of the following.)

   i. P(X ≤ 545)- _____

   ii. P(545 ≤ X ≤ 565) - _____

   iii. P(X ≤ 475) - _____

   iv. P(X ≥ 560) = _____

   v. P(500 ≤ X ≤ 565)= _____

   vi. P(485 ≤ X ≤ 500)= _____ √

c. Obtain and interpret the 67th percentile of this distribution.
   _____
                                                              √

d. Let X and Y denote, respectively, the revenues the charity will
   receive in each of the next two years, where X and Y are
   statistically independent normal random variables each with mean
   $525 million and standard deviation $20 million.

   i. Describe the probability distribution of T = X + Y, the
      combined revenue over the next two years._____
      _____
                                                              √

   ii. Suppose the charity has a goal of exceeding $1100 million in
       revenue during the next two years. What is the probability
       that this goal will be reached? _____

   iii. There is a .10 probability that combined revenue during the
        next two years will not exceed what dollar amount? _____√

C. EXPONENTIAL PROBABILITY DISTRIBUTIONS
(Text: Chapter 8, Section 8.3)

Review of Basic Concepts

C.1 To obtain an exponential probability distribution from the family of exponential probability distributions, we must specify the parameter of the distribution, which is denoted by _____. This parameter can be any _____ number.

C.2 The parameter of an exponential probability distribution is $\lambda = .4$. The mean of the distribution is _____, the variance is _____, and the distribution is skewed to the _____.

C.3 Given X is an exponential random variable with $\lambda = .4$, write the expression that must be solved to compute $P(X \leq 4)$. _____ Using this expression, we obtain $P(X \leq 4) =$ _____.

C.4 Calculate each of the following probabilities, given X is an exponential random variable with $\lambda = .15$.

a. $P(X \leq 7) =$ _____

b. $P(X > 5) =$ _____

c. $P(3 \leq X \leq 6) =$ _____ √

»Problem 3

The service life (in hours) of an electronic component, X, is an exponential random variable with $\lambda = 1/700$.

a. Determine the mean and standard deviation of the probability distribution of X. $E(X) =$ _____; $\sigma(X) =$ _____ √

b. Obtain each of the following probabilities.

i. Probability that the service life will be less than 140 hours. _____ √

    ii. Probability that the service life will exceed 700 hours. _____

    iii. Probability that the service life will be between 140 and 700 hours. _____ √

    c. Obtain the 97.5th percentile of this distribution. _____

Answers to Chapter Reviews and Problems

A.1 The two parameters are a, the smallest value the random variable can assume, and b, the largest value it can assume.

A.2 $f(x) = 1/(14-10) = .25$ where $10 \leq X \leq 14$; when graphed, the distribution appears as a rectangle with height $f(x) = .25$ and width $14 - 10 = 4$.

A.3 $E\{X\} = (14+10)/2 = 12$; $\sigma^2\{X\} = (14-10)^2/12 = 1.33$

A.4 $F(X) = (x-10)/(14-10)$ where $10 \leq x \leq 14$; $P(X \leq 11.5) = F(11.5) = (11.5-10)/(14-10) = .375$

Problem 1

    a. $a = 10$, $b = 20$; Thus, $f(x) = 1/(20-10) = .10$ where $10 \leq x \leq 20$.

    b. i.

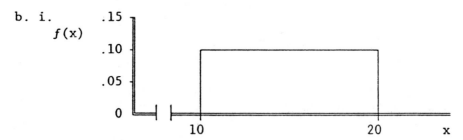

Figure 8.1

    ii. $E\{X\} = (10+20)/2 = 15$; $\sigma^2\{X\} = (20-10)^2/12 = 8.33$

    c. i. $P(X \leq 13.5) = F(13.5) = (13.5-10)/(20-10) = .35$

    ii. $P(X > 16.0) = 1 - F(16.0) = 1 - (16-10)/(20-10) = 1 - .60 = .40$

iii. $P(13.5 \leq X \leq 16.0) = F(16.0) - F(13.5) = .60 - .35 = .25$

B.1 μ; σ

B.2 No; all normal probability distributions are symmetrical (i.e., they are unskewed regardless of the values of μ and σ).

B.3 Yes; the variability of the distribution is determined by the value of σ, the standard deviation of the distribution.

B.4 Yes; the distribution is centered at μ, the mean of the distribution.

B.5 normal; mean; variance

B.6 $N(0, 1)$; Z

B.7 $P(Z \leq 1.17) = .8790$; $z(.8790) = 1.17$

B.8 .6808

B.9 .9750; .9452; .0298

B.10 $P(Z \geq 1.46) = 1 - P(Z < 1.46) = 1 - .9279 = .0721$

B.11 $P(Z \geq -1.92) = P(Z \leq 1.92) = .9726$; $P(-1.92 \leq Z \leq -1.40) = P(1.40 \leq Z \leq 1.92) = .9726 - .9192 = .0534$

B.12 $P(Z \leq -.64) = P(Z \geq .64) = 1 - P(Z < .64) = 1 - .7389 = .2611$

B.13 a. .0781; b. .0102; c. .8708; d. .3492

B.14 a. $z(.8340) = .97$

b. $z(.5000) = 0$

c. $z(.0250) = -z(.9750) = -1.96$

d. From part (b) of Table C-1, we obtain $z(.98) = 2.054$

B.15 $z = (13.5-10)/2 = 1.75$; $P(X \leq 13.5) = P(Z \leq 1.75) = .9599$

B.16 $P(9.0 \leq X \leq 13.5) = P(-.5 \leq Z \leq 1.75) = .9599 - .3085 = .6514$

B.17 a. $P(.5 \leq Z \leq 1.75) = .9599 - .6915 = .2684$

    b. $P(Z > 1.80) = 1 - .9641 = .0359$

    c. $P(Z \leq -.90) = 1 - .8159 = .1841$

    d. $P(-2.00 \leq Z \leq 2.00) = .9772 - (1 - .9772) = .9544$

    e. $P(Z > -1.25) = .8944$

    f. $P(-1.60 \leq Z \leq -1.25) = .9452 - .8944 = .0508$

B.18 normal; $\mu = 30 + 40 = 70$; $\sigma^2 = 25 + 100 = 125$

Problem 2

    a.i.

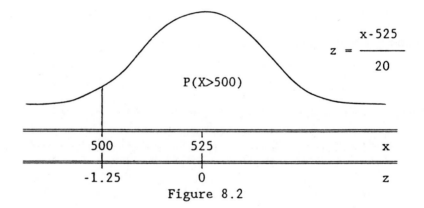

Figure 8.2

    ii. $P(X > 500) = P(Z > -1.25) = P(Z < 1.25) = .8944$

  b. i. $P(X \leq 545) = P(Z \leq 1.00) = .8413$

    ii. $P(545 \leq x \leq 565) = P(1.00 \leq Z \leq 2.00) = .9772 - .8413 = .1359$

    iii. $P(X \leq 475) = P(Z \leq -2.50) = 1 - .9938 = .0062$

    iv. $P(X \geq 560) = P(Z \geq 1.75) = 1 - .9599 = .0401$

    v. $P(500 \leq X \leq 565) = P(-1.25 \leq Z \leq 2.00) = .9772 - (1 - .8944)$
      $= .8716$

   vi. $P(485 \leq X \leq 500) = P(-2.00 \leq Z \leq -1.25) = .9772-.8944=.0828$

c. $z(.6700) = .44$; Thus, the 67th percentile of the distribution of
  X is located at the value $525 + .44(20) = 533.8$. Hence, the
  probability that the charity will realize \$533.8 million or less
  in revenue in the coming year is .67.

d. i. The probability distribution of T is normal with mean
    $\mu = 525 + 525 = \$1050$ million and the standard deviation
    $\sigma^2 = \sqrt{(20)^2 + (20)^2} = \$28.284$ million.

  ii. $P(T > 1100) = P(Z > (1100-1050)/28.284) = P(Z > 1.77) =$
    $1 - .9616 = .0384$

 iii. $z(.10) = -1.282$; The 10th percentile of the distribution of
    T is located at the value $1050 - 1.282(28.284) = 1013.74$.
    Thus, the probability that revenue during the next two years
    will be \$1,013.74 million or less is .10.

C.1 λ; positive

C.2 $E\{X\} = 1/.4 = 2.5$; $\sigma^2\{X\} = 1/(.4^2) = 6.25$; right

C.3 $P(X \leq 4) = F(4) = 1 - \exp(-1.6)$; Since $\lambda x = .4(4) = 1.60$, we obtain
    $P(X \leq 4) = .7981$

C.4 a. $P(X \leq 7) = .6501$

   b. $P(X > 5) = 1 - .5276 = .4724$

   c. $P(3 \leq X \leq 6) = -5934 - .3624 = .2310$

Problem 3

a. $E\{X\} = \dfrac{1}{1/700} = 700$; $\sigma\{X\} = \dfrac{1}{1/700} = 700$

b. i. $\lambda x = (1/700)(140) = 140/700 = .20$; Thus, $P(X < 140) = .1813$

ii. $\lambda x = (1/700)(700) = 1.00$; Thus, $P(X > 700) = 1 - .6321 = .3679$

iii. $P(140 \leq X \leq 700) = .6321 - .1813 = .4508$

c. The cell entry corresponding to .9750 is $\lambda x = 3.69$. Since $\lambda = 1/700$, we obtain $(1/700)x = 3.69$ or $x = 3.69(700) = 2583$. Thus, the 97.5th percentile is 2,583 hours.

9

STATISTICAL SAMPLING

A. POPULATION, CENSUSES, AND SAMPLES
(Text: Chapter 9, Sections 9.1 and 9.2)

Review of Basic Concepts

A.1 The total set of elements of interest in any problem is referred to as a(n) _____ .

A.2 The number of elements in a finite population is denoted by _____ .

A.3 The mean, variance, and standard deviation of a population are denoted by the symbols _____ , _____ , and _____ respectively.

A.4 Indicate whether each of the following is associated with a finite or infinite population.

    a. Process of automobile assembly in a Detroit plant. _____

    b. Process of daily demand for lawnmowers. _____

    c. Employees in an automobile assembly plant at the last year-end. _____

A.5 The six doctors in a clinic constitute a finite population with 14, 8, 7, 16, 3 and 36 years of experience, respectively. The mean years of experience in this population is _____ .

A.6 For the situation in A.5, the population variance is _____ and the population standard deviation is _____ .

A.7 The characteristic of interest in a certain infinite population can be described by a probability distribution with expected value of 12 and variance of 4. The mean of this population is _____ and the standard deviation is _____ .

A.8 Briefly indicate a major reason why a sample would be preferable to a census in each of the following situations.

   a. A study is designed by the Defense Department to examine the deterioration of M16 ammunition in storage. _____

   _____

   b. At an October 15 executive meeting, plans are made for a nationwide survey of retailer reaction to the price of a new Christmas novelty. _____

   _____

   c. A statistical analysis is designed to determine the quality of spark plugs produced in an automated assembly line. _____

A.9 Indicate whether each of the following is a probability or judgement sample.

   a. Two cities are selected by an advertising manager to test alternative advertising campaigns because they are seen as very similar cities by the manager. _____

   b. One of the six students in a graduate seminar is selected by a roll of a die to orally present a case. _____

   c. Two of the eight regional offices of a CPA firm are selected for a special program in such a manner that each of the eight offices has a one-eighth chance of being selected first, and each of the remaining seven offices has a one-seventh chance of being selected second. _____

A.10 Statistical methods can be used to assess the margin of error due to sampling only if the sample is a(n) _____ sample.

A.11 A college senior studying student attitudes toward the college newspaper gives out a questionnaire to all residents of his dormitory floor that happen to be home one evening. Explain why this selection procedure results in neither a probability sample nor a judgement sample. _____

A.12 In sampling public opinion, pollsters often require interviewers to select numbers of respondents which will make the sample representative of the population in terms of age, sex, and so forth, but leave the actual selection to the interviewers. This type of sample is called a(n) _____ sample. √

B. SIMPLE RANDOM SAMPLING
(Text: Chapter 9, Sections 9.3 to 9.5)

Review of Basic Concepts

B.1 A simple random sample from a finite population is defined as one in which each possible _____ _____ has equal probability of being chosen.

B.2 Is a simple random sample a probability or judgement sample? _____ _____ Explain. _____ _____

B.3 In most situations in which a finite population is sampled, a population element is permitted to enter the sample only one. This is referred to as sampling _____ _____.

B.4 Assume (8.9) in your Text is being used to select a simple random sample of n = 10 elements from a population of N = 200 elements. In this instance, if five sample observations have been selected, the probability of anyone of the remaining population elements being chosen as the sixth sample observation is _____.

B.5 A listing of all elements in the population is called a(n) _____.

B.6 The residential listing in the white pages of a metropolitan area telephone directory is being used as a frame to select a simple random sample from the population of all metropolitan residences. The population of all metropolitan residences with listed telephone numbers is the _____ population while the population of all metropolitan residences is the _____ population.

B.7 Observations generated by a process constitute a simple random sample from an infinite population if, prior to their generation, the observations are _____ and _____ distributed random variables.

B.8 Summary measures, such as means and variances, are referred to as _____ when calculated for a population whereas they are referred to as _____ when calculated for a sample. √

»Problem 1

There were 72 industrial companies with sales of $250 million or more last fiscal year headquartered in a certain Federal Reserve District. Earnings per share (EPS) last fiscal year for this population of N = 72 companies appear in Table 8.1. The mean and variance of this population are $\mu = 3.350$ and $\sigma^2 = 3.769$, respectively.

Table 8.1
EARNINGS PER SHARE (EPS) LAST FISCAL YEAR
FOR POPULATION OF 72 INDUSTRIAL COMPANIES

Company	EPS	Company	EPS	Company	EPS
1	$3.51	25	$2.83	49	$6.87
2	2.00	26	1.21	50	4.37
3	8.00	27	2.52	51	3.44
4	4.47	28	2.24	52	1.40
5	4.81	29	3.17	53	2.36
6	4.24	30	3.83	54	2.00
7	2.14	31	2.71	55	5.04
8	2.37	32	2.65	56	2.68
9	4.98	33	6.28	57	3.23
10	3.72	34	4.01	58	6.27
11	.84	35	1.51	59	11.04
12	4.13	36	.05	60	1.89
13	8.46	37	4.26	61	6.06
14	1.04	38	4.57	62	2.91
15	1.93	39	3.06	63	3.64
16	1.40	40	2.70	64	2.35
17	2.44	41	3.14	65	.41
18	2.73	42	2.45	66	1.92
19	4.12	43	2.46	67	2.13
20	2.22	44	5.84	68	4.41
21	3.95	45	4.00	69	3.82
22	5.55	46	2.32	70	.11
23	.80	47	1.07	71	3.12
24	3.51	48	5.29	72	2.20

a. Select a simple random sample of n = 10 companies from this population and record the EPS of each selected company. (Obtain the required random numbers from columns 21 - 25 in Table C-8 in your Text. Beginning in row 1, use the first two digits in each row as the two-digit number, and read downward.) √

Sample Element	Random Number	EPS	Sample Element	Random Number	EPS
1	24	3.51	6	____	____
2	____	____	7	____	____
3	____	____	8	____	____
4	____	____	9	____	____
5	____	____	10	____	____

b. Compute the following sample statistics for the EPS numbers in your simple random sample.

   i. $\overline{X}$ = _____

  ii. s^2 = _____

 iii. s = _____

c. The sample mean computed in part (b.i) differs from the population mean μ = 3.350. Did you expect that these means would differ? _____ Why? _____
_____ √

Answers to Chapter Reviews and Problems

A.1 population or universe

A.2 N

A.3 μ; σ^2; σ

A.4 a.infinite; b.infinite; c.finite

A.5 $\mu = 84/6 = 14.0$

A.6 $\sigma^2 = \dfrac{(14 - 14)^2 + \ldots + (36 - 14)^2}{6} = \dfrac{694}{6} = 115.67; \sigma = 10.755$

A.7 $\mu = E(X) = 12; \sigma = \sigma(X) = \sqrt{4} = 2$

A.8 a. Since the measurement of deterioration will involve expending the ammunition, a census would destroy all ammunition in storage.

   b. Since the population is large and there is little time to complete the survey before the Christmas season, it is highly unlikely that a census could be completed in time.

   c. Since the spark plug production process represents an infinite population, a census is simply not possible. (Destruction may also be a factor here if the measurement of quality renders a spark plug useless.)

A.9 a. judgement; b. probability; c. probability

A.10 probability

A.11 The elements chosen for this study were not selected according to known probabilities, as in a probability sample, nor was the selection based on their representiveness of the population, as in a judgement sample. Rather they were selected because they were conveniently available. Hence, this study group is a chunk.

A.12 quota

B.1 sample combination

B.2 probability; since the selection of elements is made according to known probabilities.

B.3 without replacement

B.4 $1/195 = .00513$

B.5 frame

B.6 sampled; target

B.7 independent; identically

B.8 parameters; statistics

Problem 1

a.

Sample Element	Random Number	EPS	Sample Element	Random Number	EPS
1	24	3.51	6	60	1.89
2	66	1.92	7	46	2.32
3	07	2.14	8	58	6.27
4	18	2.73	9	37	4.26
5	32	2.65	10	01	3.51

(Note: All random numbers larger than 72 are discarded because the population consists of only 72 companies.)

b. i. $\bar{X} = 31.20/10 = 3.12$

   ii. $s^2 = [(3.51-3.12)^2 + (1.92-3.12)^2 + ... + (3.51-3.12)^2]/\ 9 = 1.828$

   iii. $s = \sqrt{1.828} = 1.352$

c. Yes. There is no reason to expect that a sample mean $\bar{X}$ should be identical to the mean of the population from which the sample is selected. (Note: There is, however, a relationship between the sample mean X and the corresponding population mean μ which is discussed in Chapter 10 of your Text.)

A. BEHAVIOR OF $\bar{X}$ BY EXPERIMENTATION AND THEORY

(Text: Chapter 10, Sections 10.1 - 10.3)

Review of Basic Concepts

A.1 The difference between the mean obtained from a random sample and the mean obtained from a census of the population is known as

_____ _____.

A.2 Prior to sampling, the sample mean $\bar{X}$ is a(n) _____
_____, since selecting a simple random sample constitutes a random selection process. Also, after the simple random sample has been selected, the observed sample mean is a(n) _____ of a random trial of the process.

A.3 The theoretical results concerning the behavior of the random variable X presented in Section 10.3 of your Text are applicable if the population is infinite or, in the case of a finite population, if the _____ size is small relative to the _____ size.

A.4 Answer each of the following questions concerning the probability distribution associated with the random variable $\bar{X}$.

a. This probability distribution is called the _____ of $\bar{X}$.

b. The expected value of this distribution is equal to the _____ mean. Symbolically, this is written as E(X) = _____.

c. The standard deviation of this distribution is called the _____ of the mean and is denoted by _____.

d. Explain in words how $\sigma(\bar{X})$ is computed from knowledge of the sample size n and the population standard deviation σ _____

_____

    e. lf_the sample size is quadrupled, say from n = 50 to n = 200, $\sigma(\bar{X})$ will be reduced by _____ .

A.5 The Central Limit Theorem states that, for almost all populations, the sampling distribution of $\bar{X}$ is _____ _____ when the simple random sample size is sufficiently large.

A.6 A simple random sample of size n = 100 is to _be selected from a population with μ = 140 and σ = 25. Thus, $E(\bar{X})$ = _____ and $\sigma(\bar{X})$ = _____ .

A.7 For the situation in A.6, the population is known to be nearly symmetrical as well as unimodal. Thus, the sampling distribution of $\bar{X}$ with n = 100 _____ _____ √

B. PROBABILITY STATEMENTS ABOUT BEHAVIOR OF $\bar{X}$ USING NORMAL APPROXIMATION
(Text: Chapter 10, Section 10.3)

Review of Basic Concepts

B.1 Explain in words how the standardized variable is computed when dealing with the random variable $\bar{X}$. _____

B.2 A simple random sample of n = 100 is to be selected from a population with μ = 43.0 and σ = 10.0. The probability that $\bar{X}$ will be 41.5 or less is desired. Using the Central_Limit Theorem, it is anticipated that the sampling distribution of $\bar{X}$ is approximately normal when the sample size is n = 100._ Thus, the value of the standardized variable corresponding to $\bar{X}$ = 41.5 is _____ .

B.3 For the situation in B.2, $P(\bar{X} \leq 41.5)$ = $P(Z \leq$ _____) = Z _____ .

B.4 For the situation in B.2 and B.3, suppose the_simple random sample size had been n = 400. In this situation, $P(\bar{X} \leq 41.5)$ = $P(Z \leq$ _____) = _____ .

B.5 A simple random sample of size n = 64 is to be selected from a population with σ = 6.0 units. The probability that $\bar{X}$ will be within 1.5 units of the unknown population mean μ is desired. Assuming the sampling distribution of $\bar{X}$ is approximately normal when n = 64, the appropriate z values for the probability calculation are _____ and _____ .

B.6 Increasing the simple random sample size will _____ the probability that $\bar{X}$ will be within a fixed number of units from the population mean. √

»Problem 1

The infinite population of claim sizes resulting from the random process of claims filed under a certain fire insurance policy is known to have a mean of $1,250 and a standard deviation of $1,196. There is a backlog of 676 claims to be examined. Assume these 676 claims constitute a simple random sample from the infinite population of claim sizes and that the sampling distribution of X is approximately normal when n = 676.

a. What is the probability that the mean claim size $\bar{X}$ in the backlog exceeds $1,350? _____ √

b. What is the probability that the mean claim size $\bar{X}$ in the backlog is between $4200 and $4300? _____ √

C. EXACT SAMPLING DISTRIBUTION OF $\bar{X}$
(Text: Chapter 10, Section 10.4)

»Problem 2

The daily toll revenue collected at a toll bridge, $\bar{X}$, is a normal random variable with expected value $E(X) = \mu = \$12,200$, and standard deviation $\sigma(X) = \sigma = \$1400$. Toll revenues for a random sample of n = 49 days will be obtained and $\bar{X}$, the mean daily revenue, will be calculated.

a. What is the functional form of the sampling distribution of $\bar{X}$ for this situation? _____ Explain. _____
   _____
   √

b. Determine each of the following probabilities.

i. $P(12,000 \leq \bar{X} \leq 12,400)$ = _____

ii. P(11,800 ≤ $\overline{X}$ ≤ 12,600) = _____

iii. P(11,600 ≤ $\overline{X}$ ≤ 12,800) = _____ √

    c. i. Do the probabilities obtained in part (b.) depend on the
        variability in the population of toll revenues generated by
        this random process? _____ Explain. _____
        _____ √

ii. Would the probabilities in part (b.) have been larger or smaller
    had the population been more variable; that is, if σ had been
    larger than $1,400? _____ Explain. _____

Please continue with Problem 3 for an example of the construction
of the exact sampling distribution of X by enumeration.

»**Problem 3**

A service station has advertised a new, reduced price on a
particular type of automobile tire. Because of the reduced price,
however, the tires can only be purchased in pairs. The probability
distribution for X, the number of tires purchased by any customer who
inquires about the tires, is as follows.

x:	0	2	4
P(x):	.6	.3	.1

Based on this probability distribution, $E(\overline{X}) = \mu = 1.0$ and $\sigma^2(\overline{X}) = \sigma^2 =$
1.8. Let X_1 and X_2 denote, respectively, the number of tires purchased by
each of two customers who inquire about them, where X_1 and X_2 are
independent random variables each having the preceding probability
distribution (i.e., X_1 and X_2 constitute a simple random sample of size
n = 2 from the infinite population of customers who inquire about the
tires). Interest is focused on the sample mean $\overline{X} = (X_1 + X_2)/2$.

a. Identify the nine possible sample outcomes for this situation and the sample mean X associated with each outcome. Also obtain the probability of each sample outcome, recalling that X_1 and X_2 are statistically independent.√

b. i. Using your results from part (a.), construct the exact sampling distribution of X.

ii. Compute $E(\overline{X})$ and $\sigma^2(\overline{X})$ directly from the sampling distribution in part (i.)

$E(\overline{X})$ = _____ ; $\sigma^2(\overline{X})$ = _____

iii. Verify $E(\overline{X})$ and $\sigma^2(\overline{X})$ by using (10.5) and (10.6) in your Text.

_____ √

D. POINT ESTIMATION
(Text: Chapter 10, Section 10.6)

Review of Basic Concepts

D.1 When estimating a population characteristic by a single number, the number is said to be a(n) _____ _____ of the characteristic.

D.2 The estimation procedures discussed in your Text require that the sample data be obtained from a(n) _____ sample.

D.3 In a situation where we desire to estimate a population mean (μ) with a sample mean (X), prior to sampling the sample mean is known as a(n) _____, while the specific value of X obtained from the sample is referred to as a(n) _____.

D.4 In the general discussion of point estimation in your Text, indicate what each of the following denote.

a. Θ _____

b. S _____

c. E(S) _____

d. Bias _____

D.5 The sample variance $s^2 = \sum_{i=1}^{n} (X_i - \overline{X})^2 / (n-1)$ is an unbiased estimator of the population variance σ^2 since $E(s^2) = $ _____.

D.6 A television station wished to estimate the mean TV viewing time of families in its reception area last Sunday evening. A random sample 64 families was selected and $X = 2.1$ hours computed. A point estimate of μ, the mean viewing time among all families in the reception area last Sunday evening, is _____.

D.7 For the situation in A.6, is $\overline{X}$ an unbiased estimator of μ? _____ Explain. _____

D.8 Suppose S_1 and S_2 are two unbiased estimators of the parameter Θ. If S_1 and S_2 are based on the same sample size and S_1 has greater relative efficiency than S_2 in estimating Θ, then what do we know about the relative magnitudes of the variances of the sampling distributions of S_1 and S_2? _____

D.9 What role does the sample size play in the concept of consistency as a desirable property for estimators? _____
_____

Answers to Chapter Reviews and Problems

A.1 sampling error

A.2 random variable; outcome

A.3 sample; population

A.4 a. sampling distribution

   b. population; $E(\overline{X}) = \mu$

   c. standard deviation of the sampling distribution, or standard error; $\sigma(X)$

d. $\sigma(\overline{X})$ is computed as the population standard deviation σ divided by the square root of the sample size.

e. one-half

A.5 approximately normal

A.6 $E(\overline{X}) = 140$; $\sigma(\overline{X}) = 25/\sqrt{100} = 25$

A.7 approximately normal

B.1 The difference between the specified value of $\overline{X}$ and the mean μ of the population being sampled is first computed. The difference $(X - \mu)$ is then divided by the standard error of the mean $\sigma(X)$

B.2 $\sigma(\overline{X}) = \sigma/\sqrt{n} = 10/\sqrt{100} = 1$ $z = \dfrac{\overline{X} - \mu}{\sigma(\overline{X})} = \dfrac{41.5 - 43.0}{1} = -1.50$

B.3 $P(\overline{X} \leq 41.5) = P(Z \leq -1.50) = 1 - .9332 = .0668$

B.4 $z = \dfrac{41.5 - 43.0}{10\sqrt{400}} = -3.0$; thus, $P(X \leq 41.5) = P(Z \leq -3.0)$
$= 1 - .9987 = .0013$

B.5 $z = \dfrac{(\mu - 1.5) - \mu}{6\sqrt{64}} = -2.0$; $z = \dfrac{(\mu + 1.5) - \mu}{6\sqrt{64}} = 2.0$

B.6 increase

Problem 1

a. $\mu = 1250$; $\sigma(\overline{X}) = 1196/\sqrt{676} = 46$; $z = \dfrac{1350 - 1250}{46} = 2.17$;

Thus $P(\overline{X} \geq 1350) = P(Z > 2.17) = 1 - .9850 = .015$

b. $z = \dfrac{1200 - 1250}{46} = -1.09$ and $z = \dfrac{1300 - 1250}{46} = 1.09$; Thus,

$P(1200 \leq \overline{X} \leq 1300) = P(-1.09 \leq Z \leq 1.09) = .8621 - (1-.8621) = .7242.$

Problem 2

a. The sampling distribution of $\overline{X}$ is normal since the population of toll revenues follows a normal probability distribution.

b. i. $\sigma(\overline{X}) = 1400 \ /\sqrt{49} = 200$; Thus, $P(12000 \leq \overline{X} \leq 12400) = P(-1.00 \leq Z \leq 1.00) = .8413 = .8413 -.1587 = .6826$

   ii. $P(11800 \leq \overline{X} \leq 12600) = P(-2.00 \leq Z \leq 2.00) = .9772 - .0228 = .9544$

   iii. $P(11600 \leq \overline{X} \leq 12800) = P(-3.0 \leq Z \leq 3.0) = .9987 -.0013 = .9974$

c. i. Yes; $\sigma(\overline{X})$ depends, in part, on the population standard deviation, σ.

   ii. Smaller; if σ had been larger, $\sigma(\overline{X})$ would have been larger (i.e., the sampling distribution of X would have been more spread out), hence decreasing the probability that X will fall within a fixed number of units from μ.

Problem 3

a.

Sample									
Outcome $\overline{X}_1$:	0	0	0	2	2	2	4	4	4
$\overline{X}_2$:	0	2	4	0	2	4	0	2	4
Sample Mean $\overline{X}$:	0	1	2	1	2	3	2	3	4
Probability :	.36	.18	.06	.18	.09	.03	.06	.03	.10

Note: To illustrate the probability calculation, consider the case when $X_1 = 0$ and $X_2 = 0$. Since X_1 and X_2 are independent random variables, we obtain $P(X_1=0 \cap X_2=0) = P(X_1=0)P(X_2=0) = .6(.6) = .36$

b. i.
$$\frac{\overline{X}:\quad 0\quad 1\quad 2\quad 3\quad 4}{\text{Probability:}\quad .36\quad .36\quad .21\quad .06\quad .01}$$

ii. $E\{\overline{X}\} = 0(.36) + 1(.36) + 2(.21) + 3(.06) + 4(.01) = 1.0;$
$\sigma^2(\text{X-bar}) = (0 - 1)^2(.36) + (1 - 1)^2(.36) + (2 - 1)^2(.21) +$
$(3 - 1)^2(.06) + (4 - 1)^2(.01) = .90$

iii. Using (10.5) we obtain $E\{\overline{X}\} = \mu = 1.0.$ Using (10.6) we obtain
$\sigma^2(\overline{X}) = \sigma^2/n = 1.80/2 = .90$

D.1 point estimate

D.2 simple random

D.3 estimator; estimate

D.4 a. Θ is a general symbol which denotes any population parameter or characteristic of interest.

b. S is a general symbol which denotes a sample statistic used to estimate Θ. S is a function of the valued obtained in a random sample.

c. E(S) denotes the mean of the sampling distribution of the statistic S.

d. Bias is the difference between the parameter Θ and the mean of the sampling distribution of the statistic S. That is, Bias = E(S) - Θ.

D.5 σ^2

D.6 2.1 hours

D.7 Yes, $\overline{X}$ is always an unbiased estimator of μ since $E(\overline{X}) = \mu$.

D.8 $\sigma^2(S_1) < \sigma^2(S_2)$

D.9 If an estimator is a consistent estimator of a population
characteristic, then the larger the sample size, the more likely it
is that the estimate will be close to that population
characteristic.

11

INTERVAL ESTIMATION OF POPULATION MEAN

A. INTERVAL ESTIMATION OF POPULATION MEAN

(Text: Chapter 11, Section 11.2)

Review of Basic Concepts

A.1 One limitation of point estimates is that they do not indicate how precise the estimate is. What is meant by precision of the estimate in this context? _____

_____

A.2 In a study to estimate μ, the mean number of hours of use obtained from a certain type of transistor, results from a random sample of transistors were used to construct an interval estimate for μ. The lower bound of this estimate was 346.1 hours, the upper bound was 365.3 hours. Thus, the appropriate interval estimate is _____.

A.3 What symbol is used to denote each of the following?

a. Population variance. _____

b. Variance of the sampling distribution of $\overline{X}$. _____

c. Sample variance. _____

d. Estimated variance of the sampling distribution of $\overline{X}$. _____

A.4 $s(\overline{X})$ is called the _____ _____ _____ of $\overline{X}$.
Explain in words how $s(\overline{X})$ is computed. _____

_____

A.5 What is the relationship between $s(\overline{X})$ and $\sigma(\overline{X})$? _____

_____

A.6 In constructing a 1% confidence interval for μ based on a reasonably large sample size, the value for $z(1 - \alpha/2)$ must be determined. Complete the following with the assistance of part b of Table C-1 in your Text, for the stated α levels.

α	Confidence Coefficient	(1 - α/2)	z(1 - α/z)
.20	.80	.90	1.282
.10	___	___	___
.05	___	___	___
.02	___	___	___
.01	___	___	___

A.7 A production process for pre-fabricated homes results in an unknown amount of scrap lumber each day. In a study to estimate the mean daily board feet of scrap 2x4 lumber, a_random sample of 64 days yielded the following sample results: X - 320 and s - 16 board feet. To construct a 98 percent confidence interval for μ, (1 - α/z) = _____ z(1 - α/2) - _____, and s{X} - _____.

A.8 For the situation in B.7, the confidence limits are _____ ± _____. Hence, the 98 percent confidence interval is _____.

A.9 A statistical study was conducted to estimate μ, the mean exposure time (in milliseconds) of an automatic camera lens under conditions of full sunlight. The 90 percent confidence interval which. resulted was $5.61 \le \mu \le 5.73$. Explain what this confidence interval means. _____

A.10 While it is desirable to have a narrow confidence interval with a high level of confidence, often this would require that a very _____ simple random sample be taken.

A.11 For any fixed sample size, if the confidence coefficient is increased the width of the interval estimate is _____.

A.12 The marketing research analyst for a fast food franchisor constructs all confidence intervals used in a particular type of analysis with a confidence coefficient of .90. He figures that about 2,000 such intervals have been constructed in the last ten years, each from a different (i.e., independent) random sample.

About _____ of these intervals will not have been correct and
there is a _____ probability that the next interval
constructed will not be correct. √

»Problem 1

An auditor for a CPA firm desires to estimate the mean value of
items carried in an inventory account. A random sample of 100 items
from the 8,000 items in the inventory account   yields X = $206 and s =
$63.   (Since the sampling fraction n/N = 100/8,000 = .0125 is small, the
estimators given by (11.2) in your Text are applicable.)

a. Construct a 95 percent confidence interval for μ. _____ √

b. Construct a 99 percent confidence interval for μ. _____

c. Which of the two confidence intervals above is wider? _____
   What does this imply about the relationship between the
   confidence coefficient and the precision of the estimate?
   _____
                                                                  √

d. If a second random sample of 100 items were selected from the
   original population of 8,000 items and a 95 percent confidence
   interval for μ constructed from these new sample results, would
   the confidence limits be the same as those computed in (a.)?
   _____ Explain. _____
   _____
                                                                  √

B.   OTHER CONFIDENCE INTERVALS
(Text: Chapter 11, Sections 11.3 and 11.4)

Review of Basic Concepts

B.1 If we select a random sample of size n from a normal population,
    $(X - \mu)/ s(X)$ follows a _____.distribution with
    _____ degrees of freedom.

B.2 Since the t distribution applies adequately in a wide range of
    cases where the population is not exactly normal, it is referred to
    as being _____.

B.3 From Table C-3 in your Text, determine the appropriate value of the t distribution to use in constructing a confidence interval for μ for the following levels of α and n

α	n	t(1-α)/2; n-1)
.10	20	1.729
.05	15	————
.05	20	————
.02	25	————
.01	25	————

B.4 A personnel manager wishes to estimate mean weekly earnings of the 4,200 factory workers in a community. A random sample of n = 25 workers yields X = $330 and s = $40. Past experience indicates that the distribution of weekly earnings is nearly normal. If a 95 percent confidence coefficient is specified, t(1 - α/2; n - 1) = _____, s(X) = _____ and, hence, the appropriate confidence limits are _____ ± _____.

B.5 For the situation in C.4, the resulting confidence interval is _____ Is the confidence coefficient for this interval exactly .95? _____ Explain. _____

B.6 The Greek symbol τ is used to denote the _____ _____.

B.7 For the situation in C.4, a 95 percent confidence interval for τ, the total weekly earnings among the 4,200 community factory workers, is _____ ≤ τ ≤ _____.

B.8 When $\sigma(\overline{X})$ is known exactly and the population is normal, we use values from the _____ _____ distribution in constructing confidence intervals regardless of the sample size.√

»Problem 2

a. It is desired to estimate the mean time students spend per
    session a computer terminal in a large university. A random
    sample of 16 sessions yields X = 28.0 minutes and s = 6.0
    minutes. Session times are known to follow a normal
    distribution. Construct a 99 percent confidence interval for μ.
    _____ √

b. Suppose the auditor in Problem 1 wanted to estimate the total
    value of the 8,000 inventory items. In Problem 1(a.) we
    constructed a 95 percent confidence interval for the mean value;
    193.65 ≤ μ ≤218.35. Construct a 95 percent confidence interval
    for τ. _____ √

c. The owner of a cultured sod farm wishes to determine the extent
    to which his sod has been infested with dandelions. A random
    sample of 8 plots of sod, each ten centaurs (i.e., ten square
    meters) in size, is sel. The dandelion plants found on these
    plots number 7, 3, 8, 11, 6, 4, 8, and 5, respectively.

    i. Determine $\overline{X}$ and s for this sample. $\overline{X}$ = _____
        s = _____ . √

    ii. Construct a 90 percent confidence interval for μ, the mean
         number of dandelion plants per plot. Assume the distribution
         of numbers of dandelion plants is normal. _____ √

    iii. If the farm consists of 30,000 plots of this size, obtain a
          90 percent confidence interval for τ, the total number of
          dandelion plants infesting the farm. _____ √

C. PLANNING OF SAMPLE SIZE
(Text: Chapter 11, Section 11.5)

Review of Basic Concepts

C.1 In planning the sample size, we generally need to obtain a planning
     value for the population _____ _____.

C.2 Why would it usually not be advisable to simply select an arbitrary sample size which is so large that it would meet any requirements for precision and confidence level? _____

_____

C.3 One term in (11.18) in your text is h. It represents the specified or desired margin of _____ _____ and is called the _____ of the confidence interval.

C.4 The term z in (11.18) in your Text is the standard normal percentile associated with the specified or desired _____

_____ .

C.5 If a sample is based on an arbitrarily selected sample size and the resulting confidence interval is wider than desired, then the sample size used was too _____ . √

   »Problem 3

   A study of the mean number of programmer errors per page of coding is to be undertaken in a commercial data processing division. Management intends to use the information to establish performance and time standards. It is desired to estimate this mean within ±.10 error per page with a 95.44-percent confidence coefficient. Past studies suggest that 1.2 is a suitable planning value for the population standard deviation.

   a. Determine the required sample. (Note: $z = z(1 - \alpha/2) = z(.9772)$ = 2.0.) _____ √

   b. Suppose a random sample of 576 pages of coding is selected and yields $\bar{X} = 3.5$ and $s = 1.6$.

      i. Construct the 95.44 percent confidence interval for μ.
         _____

      ii. Did this confidence interval provide the desired precision? _____ Why? _____ √

   c. What would the required sample size in part (a.) have been had management specified a 99.74 percent confidence coefficient? (Note: $z = z(1 - \alpha/2) = z(.9987) = 3.0$.) _____

d. What do your answers to parts (a.) and (c.) imply about the relationship between desired confidence level and required sample size? _____ √

Answers to Chapter Reviews and Problems

A.1 Precision refers to the magnitude of error due to sampling. A very precise estimate is one that contains little sampling error. Point estimates do not indicate the amount of sampling error in the estimate.

A.2 $346.1 \leq \mu \leq 365.3$

A.3 a. σ^2

   b. $\sigma^2\{\overline{X}\}$

   c. s^2

   d. $s^2\{\overline{X}\}$

A.4 estimated standard deviation or estimated standard error; For infinite populations or finite populations in which the sampling fraction is small, s{X} is computed as the sample standard deviation divided by the square root of the sample size.

A.5 $s\{\overline{X}\}$ is an estimator of $\sigma\{\overline{X}\}$ obtained by replacing the population standard deviation σ in the formula $\sigma\{\overline{X}\} = \sigma/\sqrt{n}$ with the sample standard deviation s. (Recall that σ is generally unknown in actual applications.) Hence, $s\{\overline{X}\} = s/\sqrt{n}$.

A.6

α	Confidence Coefficient	$(1-\alpha/2)$	$z(1-\alpha/2)$
.20	.80	.90	1.282
.10	.90	.95	1.645
.05	.95	.975	1.960
.02	.98	.99	2.326
.01	.99	.995	2.576

A.7 .99; z(.99) = 2.326; s($\overline{X}$) = 16/$\sqrt{64}$ = 2.0

A.8 320 ± 2.326(2); 315.35 ≤ μ ≤ 324.65

A.9 With 90 percent confidence, we can state that the mean exposure time of the lens is between 5.61 milliseconds and 5.73 milliseconds. The confidence coefficient indicates that if this estimation procedure were repeated with many independent random samples of the same type for the study, the resulting interval estimates would be correct (i.e., would contain μ) in 90 percent of the cases.

A.10 large

A.11 increased

A.12 200; .10

Problem 1

a. z(.975) = 1.96; s($\overline{X}$) = 63/$\sqrt{100}$ = 6.3; the confidence limits are 206 ± 1.96(63). Alternatively, L – 206 - 1.96(6.3) - 193.65 and U – 206 ± 1.96(63) – 218.35. Thus, the 95% confidence interval for μ is 193.65 ≤ μ ≤ 218.35. Thus, the 99% confidence interval for μ is 189.77 ≤ μ ≤ 222.23.

b. z(.995) = 2.576; s($\overline{X}$) – 6.3. The confidence limits are 206 ± 2.576 (6.3).

c. The 99 percent confidence interval is wider than the 95 percent confidence interval. This implies that, for a given sample result, the confidence coefficient can be increased only by increasing the width and, hence, decreasing the precision of the confidence interval.

d. Not necessarily. The second sample would most likely result in a different s and X, both of which affect the confidence limits.

B.1 t; n - 1

B.2 robust

B.3

α	n	$t(1 - \alpha/2; n - 1)$
.10	20	1.729
.05	15	2.145
.05	20	2.093
.02	25	2.492
.01	25	2.797

B.4 $t(.975; 24) = 2.064$; $s(\overline{X}) = 8$; $330 \pm 2.064(8)$

B.5 $313.49 \leq \mu \leq 346.51$; No; Since the population is not exactly normal, the confidence coefficient is only approximately .95.

B.6 population total

B.7 $1,316,658 = 4,200(313.49) \leq \tau \leq 4,200(346.51) = 1,455,342$

B.8 standard normal

Problem 2

   a. $t(.995;15) = 2.947$; $s(\overline{X}) = 6.0/4 = 1.5$;   The confidence limits are $28.0 \pm 2.947(1.5)$.   Thus, the 99 percent confidence interval for μ is $23.580 \leq \mu \leq 32.420$.

   b. $1,549,200 = 8,000(193.65) \leq \tau \leq 8,000(218.35) = 1,746,800$

   c. i. $\overline{X} = 52/8 = 6.5$; $s = \sqrt{46/7} = 2.563$

   ii. $t(.95; 7) = 1.895$; $s(\overline{X}) = 2.563/\sqrt{8} = .906$; The confidence limits are $6.5 \pm 1.895(.906)$. Thus, the 90 percent confidence interval for μ is $4.783 \leq \mu \leq 8.217$.

   iii. $143,490 = 30,000(4.783) \leq \tau \leq 30,000(8.217) = 246,510$

C.1 standard deviation

C.2 Unnecessary expense would be incurred by continually selecting larger samples than needed.

C.3   sampling error; half-width

C.4   confidence level or confidence coefficient

C.5   small

Problem 3

    a. $z(.9772) = 2$; $\sigma = 1.2$; $h = .1$; Thus, $n = (2)^2(1.2)^2/(.1)^2 = 576$ pages.

    b.   i. $z(.9772) = 2$; $s(\bar{X}) = .067$; Thus, the 95.44 percent confidence interval for μ is $3.366 \le \mu \le 3.634$.

         ii.   No. The actual half-width of the interval, $2(.067) = .134$, is somewhat larger than desired. This occurred because the sample standard deviation used in constructing the confidence interval, $s = 1.6$, is somewhat larger than 1.2, the planning value for σ.

    c. $n = (3)^2(1.2)^2/(.1)^2 = 1,296$ pages

    d.   The required sample size increases as the specified confidence level increases.

12

TESTS FOR POPULATION MEAN

A. STATISTICAL TESTS AND NATURE OF STATISTICAL DECISION RULES
(Text: Chapter 12, Sections 12.1)

Review of Basic Concepts

A.1 In statistical tests about a population mean, one-sided upper-tail alternatives are of the form H_0: μ_____μ_0 and H_1: μ_____μ_0, while two-sided alternatives are of the form H_0: μ_____μ_0 and H_1: μ_____μ_0.

A.2 For the situation in A.1, what does the symbol μ_0 denote? _____
_____

A.3 The alternative conclusions H_0 and H_1 in a statistical test are often called hypotheses. Specifically, H_0 is referred to as the _____ hypothesis of the test.

A.4 Why would the alternative conclusions or hypotheses H_0: $\mu \geq 700$ and H_1: $\mu < 800$ be inappropriate in a statistical test? _____
_____

A.5 For any statistical test based on sample data, two types of errors of inference can be made. A Type I error is made if _____ is selected correct when, in fact, _____ is correct. A Type II error is made if _____ is selected as correct, when, in fact, _____ is correct.

A.6 The probability or risk of making a Type I error is called a(n) _____ risk and the probability or risk of making a Type II error is called a(n) _____ risk.

A.7 A sales contract specifies that μ, the mean height of seedling shrubs delivered at retail, is 12.0 inches. In a statistical test involving μ, it is desired to control the risk of concluding the mean height is less than 12.0 inches when, in fact, it is 12.0 inches or greater. If this is to be the α risk, state the appropriate alternatives. _____

_____

A.8 For the situation in A.7, suppose the retailer is instead concerned about incorrectly concluding the mean height is greater than 10.0 inches. If this is to be the α risk, state the appropriate alternatives. _____

A.9 An agricultural agent plans to assess the impact of a recent drought on wheat fields in a farming region. If the mean yield per acre is sufficiently low, the agent can apply for government assistance for the region. The agent is primarily concerned with guarding against the error of failing to apply for assistance for the region if the mean yield is 30 bushels or less per acre. Specify the alternatives for this one-sided statistical test in such a manner that the primary concern of the agent becomes the Type I error. H_0:_____; H_1:_____

A.10 For the situation in A.9, describe what is meant by a Type I error and a Type II error.

Type I error. _____

Type II error. _____

Does this description imply that, regardless of the sample data, one of these two errors must occur? _____ Explain.

A.11 An electronic chip supplier claims that its product has a mean life of at least 200 hours. A firm that requires a large number of these chips wishes to test this claim. If the claim is correct, the firm will buy the product. If not, an alternate supplier of higher cost chips with known reliability will be used. The firm's purchasing agent is concerned with controlling the rather expensive error of unnecessarily using the alternate supplier; that is, of incorrectly concluding that the first supplier's claim is false. Specify the alternatives for this test so that the purchasing agent's concern becomes the Type I error.

H_0: _____ H_1: _____

A.12 For the situation in A.11, assume the firm decides to use the alternate supplier based upon sample data. If, in fact, the mean service life of the electronic chips from the first supplier is 223 hours, which type of error, if any, has been made? _____ Explain. _____

A.13 For the situation in A.11, would an error have been made by the firm in using the alternate supplier if the mean service life of chips from the first supplier is 185 hours? _____ Explain.
_____

A.14 For tests involving a population mean, μ, a(n) _____ _____ specifies which alternative should be selected for each possible outcome of the test statistic, X.

A.15 Consider the following statistical decision rule for a one-sided lower-tail test involving μ:

If $\bar{X} > 22$, conclude $H_0(\mu \geq 25)$ If $\bar{X} < 22$, conclude $H_1(\mu < 25)$

The value 22 here is called the _____ _____ of the decision rule, which is in general denoted _____. The (as yet undetermined) sample mean is called the _____.

A.16 For the situation in A.15, identify the range of values of the test statistic X that comprises the acceptance region and the range that comprises the rejection region. _____

A.17 Your Text states that if the alternative conclusions in a statistical test are formulated so the α risk is the more serious risk, H_1 should be concluded only if the sample evidence clearly suggest that H_1 is true. For each of the following tests, then, identify in general terms where the action limit(s) of the associated decision rule should be set with respect to the standard μo in the alternatives.

a. H_0: $\mu \leq 1,000$ and H_1: $\mu > 1,000$ _____
b. H_0: $\mu = 4,500$ and H_1: $\mu \neq 4,500$ _____
c. H_0: $\mu \geq 2,700$ and H_1: $\mu < 2,700$ _____ √

B. NATURE OF STATISTICAL DECISION RULES
(Text: Chapter 12, Section 12.2)

Review of Basic Concepts

B.1 Consider the alternatives H_0: $\mu \leq 30$ and H_1: $\mu > 30$ in a one-sided upper-tail test. The decision maker here wants just a 1 in 10 chance of concluding H_1 if the population mean is exactly 30. Symbolically, this risk specification is written $\alpha =$ _____ when $\mu = \mu o =$ _____ .

B.2 Suppose the alternatives in a test are H_0: $\mu \leq 65.0$ and H_1: $\mu > 65.0$. A simple random sample of n = 100 items from the population of interest yields X = 69.3 and s = 15.0. Describe the relevant sampling distribution of the X test statistic for the risk specification α = .05 at $\mu = \mu o = 65.0$.

a. Shape: _____

b. Mean: _____

c. Estimated Standard Deviation: _____

B.3 For the situation in B.2, (X-65.0)/1.5 has an approximate standard normal distribution. Therefore, it is called the _____ test statistic, and is denoted by _____ .

B.4 For the situation in B.2, the action limit A must correspond to the _____ th percentile of the standard normal distribution. Thus, the appropriate decision rule stated in terms of the standardized test statistic is: If _____ conclude $H_0(\mu \leq 65.0)$; If _____ conclude $H_1(\mu > 65.0)$.

B.5 Assume a statistical test with risk specification α = .10 at μ_0. If the alternatives in this test happen to be of the one-sided upper-tail form, the boundary between the acceptance and rejection regions occurs at the position _____ on the z* scale. On the other hand, if the alternatives are of the one-sided lower-tail form, the boundary occurs at the position _____ on the z* scale. Explain why this boundary differs between upper-tail and lower-tail tests. _____

B.6 For a test involving the alternatives H_0: $\mu \leq 45.0$ and H_1: $\mu > 45.0$ with sample results n - 144, X - 47.3 and s - 12.0, the position of the action limit was A = 47.054. For this test, the α risk when μ = μo = 45.0 must have been established at what level? _____ Explain. _____

B.7 In selecting between the alternatives H_0: $\mu \geq 200$, and H_1: $\mu < 200$, the following sample results were obtained: n - 400, X - 192 and s = 50. For the risk specification α - .05 when μ - μo - 200, state the appropriate decision rule in terms of the standardized test statistic. _____

B.8 For the situation in B.7, $s(\overline{X})$ - _____ and z* - _____. Hence, what conclusion should be reached? _____ Explain. _____

B.9 For the situation in B.7, the position of the action limit for the test statistic X is A - _____.

B.10 In a two-sided test involving μ, alternative H_1 should be concluded only if the test statistic X departs markedly from μ_0 in either direction, or, alternatively, only if the standardized test statistic z* departs markedly from zero in either direction. Another way of stating this is that alternative H_1 should be concluded only if the absolute value of z* is sufficiently

_____.

B.11 Consider the alternatives H_0: $\mu = 12.5$ and H_1: $\mu \neq 12.5$ in a two-sided test with the risk specification α = .01 when μ - μo = 12.5. In this test the boundaries between the acceptance and rejection regions occur at the positions _____ and _____ on the z* scale. Hence, the appropriate decision rule is _____.

B.12 For the situation in B.11, a random sample of n = 49 yields X - 16.2 and s - 10.5. Given these sample results, $s(X)$ - _____ and z* - _____. Hence, what conclusion should be reached? _____ _____ Explain. _____

_____ √

»Problem 1

A food grower sends large shipments of potatoes to processors. If the mean percent of sugar content in the potatoes in any shipment is less than or equal to 15.0, the shipment is of acceptable quality and the regular price is received. If the mean content exceeds 15.0, the shipment is considered to be of lower quality and a lower price is received. The grower takes a random sample of n = 36 potatoes from an outgoing shipment. The sample results are X = 16.4 and s = 3.6. The following risk specification has been determined: α = .10, when $\mu = \mu_0 = 15.0$.

a. Specify the alternatives for this problem. _____

_____ √

b. i. Construct a diagram similar to Figure 11.2 in your Text to illustrate the decision rule for this problem.

   ii. State the resulting decision rule. _____

   _____ _____ √

c. Determine the value of z*, the standardized test statistic, for this problem. _____

_____ √

d. What conclusion should be reached? _____

»Problem 2

Miracle Manufacturing is considering a new procedure for assembling units. The mean assembly time using the old procedure is 16 minutes. Management wishes to know whether or not the mean assembly time μ for the new procedure is less than that for the old one. In a random sample of 64 assembly times with the new procedure, X = 15.07 and s = 4.0. Management has selected a risk specification of α = .01 when $\mu = \mu_0 = 16.0$.

a. Specify the alternatives for this problem. _____

b. i. Construct a diagram similar to Figure 11.3 in your Text to illustrate the decision rule for this problem.

   ii. State the resulting decision rule. _____

   _____ √

c. Determine the value of z* for this problem.

d. Based on your decision rule, should the new procedure be
    adopted? _____ Explain. _____
    _____ √

A restaurant food distributor has a contract with a western "fish
farm" which states that the mean length of delivered trout should
exactly equal 13.0 inches. To check adherence to the contract
requirements, the distributor selects a random sample of n = 100
delivered trout and measures each. The sample results are X =
12.91 and s = .90. The α risk is to be controlled at the .10 level
when $\mu - \mu_0 = 13.0$.

a. Specify H_0 and H_1 for this problem. _____

b. State the appropriate decision rule. _____
    _____ √

c. Determine the value of z* for this problem. _____

d. Should the distributor conclude that the contract requirement is
    being met? _____ Explain. _____
    _____ √

C. P-VALUES
(Text: Chapter 12, Section 12.3)

Review of Basic Concepts

C.1 In a statistical test involving the alternatives H_0: $\mu \leq 45$ and H_1:
     $\mu > 45$, the sample results were   n = 100,   X = 48.0 and s = 18.0.
     Hence, z* = 1.67 and the P-value = .0475. Explain what the P-value
     indicates in this situation. _____
     _____

C.2 For any statistical test involving a population mean, the smaller
     the P-value is (i.e., the closer it is to zero) the _____
     consistent the sample evidence is with alternative H_0 being true.

C.3 Separate statistical tests involving the alternatives $H_0: \mu \geq 150$ and $H_1: \mu < 150$ were conducted on the means of two different populations using the same sample size and n=300. The resulting P-values were .4753 for population 1 and .0076 for population 2. For which population was the respective sample evidence more consistent with conclusion $H_0 (\mu \geq 150)$ being true? _____ Explain. _____

C.4 For the situation in C.1, suppose the n risk is specified at the .01 level when $\mu - \mu_0 - 45.0$. Given the P-value - .0475, alternative _____ is concluded.

C.5 In selecting between the alternatives $H_0: \mu \geq 200$ and $H_1: \mu < 200$, the sample results n - 400, X - 192, and s - 50 lead to the standardized test statistic z* - -3.20. Hence, the P-value for this test is computed as P(Z _____ z*) - P(Z _____ -3.20 since we are interested in the probability that z* could have been _____ than (i.e., more extreme than) the observed outcome z* - -3.20.

C.6 For the situation in C.5, the P-value - _____. Given α - .05 is specified when $\mu - \mu_0 - 200$, alternative _____ is concluded.

C.7 In selecting between the alternatives $H_0: \mu - 10$ and $H_1: \mu \neq 10$, the sample results n = 81, X - 11.2, and s - 4.5 lead to the standardized test statistic z* - 2.40. Hence, p(Z > z*) - P(Z > 2.40)- - -0082. Explain in words why the two-sided P-value for this test is 2(.0082) - .0164. _____

C.8 Determine the P-value for each of the following statistical tests from problems in section B.

	Alternatives	n	$\overline{X}$	$s(\overline{X})$	z*	P-value
a.	$H_0: \mu \leq 15$ $H_1: \mu > 15$	36	16.40	.60	2.33	_____
b.	$H_0: \mu \geq 16$ $H_1: \mu < 16$	64	15.07	.50	-1.86	_____
c.	$H_0: \mu - 13$ $H_1: \mu \neq 13$	100	12.91	.09	-1.00	_____

»Problem 4

An air freight company wished to test whether or not the mean weight of parcels shipped on a particular route exceeds ten pounds. A random sample of 49 shipping orders was examined. The results of that examination were X=9 pounds and s = 2.8 pounds. The risk specification is α = .05 when $\mu = \mu_0 = 10$. Conduct the appropriate statistical test using P-values. Specify the alternatives and decision rule, determine the P-value for the test, and state the resulting decision. √

D. SMALL SAMPLE TESTS
(Text: Chapter 12, Section 12.4)

»Problem 5

A market research class decided to evaluate an advertised claim that a particular gasoline additive yields a mean increase in mileage of more than 1.50 miles for each 10 gallons used. The class randomly selected a sample of 25 cars on which to test this claim and found a mean increase of 1.48 miles with a standard deviation of .20 in the sample. Assuming the population is normal, conduct the appropriate test, given the α risk is specified at the .10 level when $\mu = \mu_0 = 1.50$._ Specify the alternatives and decision rule, compute t* = $(X-\mu_0)/s(X)$, and state the resulting decision. √

»Problem 6

An agricultural research specialist is concerned that an insecticide promoted as being effective in controlling against infestation of corn stalks by a certain insect may not be so. It is known that the infestation rate with no treatment is normally distributed with a mean of 30 infested stalks per tract. The researcher applied the insecticide in question to a random sample of 9 tracts and obtained the following results: X = 24.5 and s = 4.5. It is desired to control the risk of incorrectly concluding that the insecticide reduces the rate of infestation when $\mu = \mu_0 = 30$ at the $\alpha = .01$ level. Assuming that the population of infested stalks per tract remains normal even if the insecticide is applied, conduct the appropriate test for μ. Specify the alternatives and decision rule, compute t*, and state the resulting decision. √

»Problem 7

Trucking industry standards specify that the time required for a certain short-run haul should equal 60 minutes. One trucking company wishes to test whether its fleet is performing at industry standards for this haul. Sixteen randomly selected hauls are timed with the following results: X = 71.23 and s = 14.5. Conduct the appropriate test for μ, given a risk specification of α = .05 when μ = μ_0 = 60 and assuming that the distribution of times over this haul by the company fleet does not depart markedly from a normal distribution. Specify the alternatives and decision rule, compute t*, and state the resulting decision. √

E. POWER OF TEST
(Text: Chapter 12, Section 12.6)

Review of Basic Concepts

E.1 In selecting between alternatives H_0: $\mu \le 200$ and H_1: $\mu > 200$, an action limit of A = 215 has been determined. State the appropriate decision rule for this situation in terms of the test statistic X.

E.2 For the situation in E.1, explain in words what is denoted by $P(H_1; \mu = 200)$. _____

_____

E.3 For the situation in E.1, suppose that $P(H_1; \mu = 220)$ =.7123. The probability .7123 is called the _____ probability of the decision rule at $\mu = 220$, and the graph of $P(H_1; \mu)$ for different possible values of μ is called the _____

_____ _____.

E.4 For the situation in E.1, if $P(H_1; \mu = 220)$ = .7123, then the ß risk at $\mu = 220$ is _____.

E.5 For the situation in E.1. suppose n and σ are such that the following rejection probability curve is obtained.

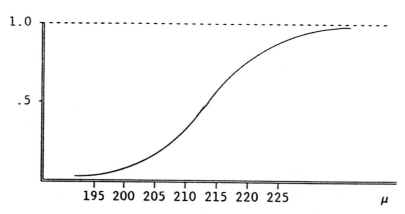

Figure 11.1

a. Which portion of this curve measures α risks? _____
   _____

b. Which portion of this curve measures ß risks? _____
   _____

c. Reading from the graph, approximately what is the risk of error
   if μ = 220? _____ Is this an α risk or a ß
   risk?

d. Is there an α risk if μ = 210? _____ Explain. _____
   _____

e. Reading from the graph, approximately what is the α risk if
   μ=200? _____

E.6 Is the following statement true or false? "To evaluate the risks in
a decision rule we must compute $P(H_1; μ)$ or its complement for
values of μ because μ can vary depending upon the sample outcome.
_____ Explain. _____
_____

E.7 Consider the alternatives H_0: $\mu \leq 85$ and H_1: $\mu > 85$ where it is known that $\sigma = 25.0$ and the risk specification is $\alpha = .05$ when $\mu = \mu_0 = 85$. It is desired to conduct this test with a random sample of size 100. The appropriate action limit for the test statistic X is A = μ_0 + Z(1 - α)σ(X) = _____.

E.8 For the situation in E.7, P(H_1; $\mu = 95$) = _____. What is the ß risk at $\mu = 95$? _____

E.9 For the situation in E.7, P(H_1; $\mu = 84$) = _____. What is the ß risk at $\mu = 84$? _____

E.10 Consider the following rejection probability curve.

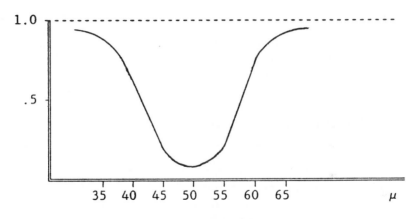

Figure 1↑.2

a. Judging by the shape of this curve and its placement along the μ axis, specify what the alternative conclusions associated with this curve must have been. _____

_____

b. Reading from the graph, what is the risk of error if $\mu = 50$? _____ Is this an α risk or a ß risk? _____

c. Reading from the graph, approximately what is the risk of error if $\mu = 40$? _____ Is this an α risk or a ß risk? _____

E.11 Given that the sample size decision rule that decreases the α risk at μ_0 will _____ the ß risks at values of μ where H_1 is true. √

»**Problem 8**

A particular model of power lawn mower yields a mean of 26 hours of use per gallon of gasoline, according to company tests. A new, less expensive motor design for the mower has been developed and the designers claim mean hours of use with the new motor at least equals that of the original design. Management wishes to test this claim against the alternative that mean hours of use with the new motor is less than 26 hours per gallon. If it does equal 26 hours or more per gallon, they plan to adopt the new motor for this mower.

a. Specify the alternatives and associated management actions for this statistical test. _____

_____

b. Suppose that management has a random sample of n = 49 mowers with the new motor design built and tested. It can be assumed from past experience with motors of this kind that σ = 2.8. Hence, with a sample of 49, $\sigma(\bar{X})$ = 2.8/$\sqrt{49}$ = .4. The risk specification is α = .05 when μ = μ_0 = 26. Specify the decision rule for this test in terms of the test statistic X. _____

_____

_____ √

c. i. Determine the rejection probability if the mean hours of use per gallon with the new motor is actually μ = 25.0. (Note: You will need to draw a figure analogous to Figure 12.10(b) in your Text but which reflects that this is a lower-tail test.)˙√

   ii. Using the same procedure, determine P(H$_1$; μ = 25.5).

  iii. Continuing with this procedure, determine P(H$_1$; μ) for the remaining μ values given in the following table. √

μ	P(H$_1$;μ)		μ	P(H$_1$;μ)
24.75	_____		25.75	_____
25.00	_____		26.00	.0500
25.25	_____		26.25	_____
25.50	_____		26.50	_____

d. i. Using your results from part (c.), sketch the rejection probability curve for this decision rule.

  ii. Using the curve, label the α risk if $\mu = 26.25$.

 iii. Using the curve, label the ß risk if $\mu = 25.00$.

  iv. Explain why this curve tends to fall from left to right while the rejection probability curve shown in Figure 12.11 of your Text tends to rise from the left to right. _____
_____ √

»Problem 9

The director of the Employment Services Department of a large metropolitan area desired to test whether or not the mean change in annual family income between the last two years, denoted by μ, was zero. A random sample of n = 2,500 area families was selected and, for each family, the change in annual family income between the two years was measured.

a. Specify H_0 and H_1 for this problem. _____

b. Define a Type I and a Type II error in the context of this problem.

  i. Type I error. _____
     _____

 ii. Type II error. _____
     _____ √

c. The director specified that the risk of incorrectly concluding there has been a change in mean family income be controlled at the .05 level. Specify the decision rule for this test in terms of the test statistic X. (Assume that the population of families is very large relative to the sample size and that the population standard deviation is $\sigma = \$1,000$.) _____
_____

d. i. Determine the rejection probability if mean annual family income has decreased by $10; that is, when $\mu = -10$. (Note: You will need to draw a figure similar to Figure 12.13(b) in your Text.)

ii. Determine the rejection probability if mean annual family income has increased by $10; that is, when $\mu = 10$.

iii. What generalization is suggested by the results in (i.) and (ii.)? _____

_____

e. i. Construct the rejection probability curve for this decision rule. Use ordinates corresponding to $\mu = -70, -50, -30, -10, 0, 10, 30, 50,$ and 70 in constructing your curve.

ii. Does the specified decision rule result in substantial ß risks if, in fact, there has been a slight change in mean income? _____ Explain. _____

_____

iii. Does the specified decision rule result in substantial ß risks, if, in fact, there has been a large change, say $\mu = 70$ or $\mu = -70$, in mean income? _____ Explain. _____

_____

F. PLANNING OF SAMPLE SIZE
(Text: Chapter 12, Section 12.7)

Review of Basic Concepts

F.1 In calculating the sample size necessary to control both α and ß risks, a planning value for _____ must be used.

F.2 In upper-tail tests involving μ, the ß risk is specified at some value of μ _____ than μ_0. The value of μ at which the ß risk is specified is denoted by _____.

F.3 The alternatives in a test are H_0: $\mu \leq 7.5$ and H_1: $\mu > 7.5$. It is desired to control the probability of a Type I error at the .025 level when the population mean is 7.5. Also, it is desired to control the probability of a Type II error at the .05 level when the population mean is 8.0. Write both of these risk specifications symbolically, using the notation for rejection probabilities. _____

F.4 For the situation in F.3, $z_0 =$ _____ and $z_1 =$ _____.

F.5 For the situation in F.3, substitute into (11.20) in your Text to obtain the required sample size, given α = .90 is a reasonable planning value.

F.6 Consider a lower-tail test involving the alternatives H_0: $\mu \geq 50$ and H_1: $\mu < 50$. The α risk is to be controlled at the .01 level when $\mu = \mu_0 = 50$ and the β risk is to be controlled at the .05 level when $\mu = \mu_1 = \underline{45}$. Therefore, in planning the sample size, the interval on the X scale from $\mu = \mu_1 = 45$ to the action limit A must equal _____ standard deviations, the interval from A to $\mu = \mu_0 = 50$ must equal _____ standard deviations and, hence, the whole interval from μ_1 to μ_0 must equal _____ standard deviations.

F.7 For the situation in F.6, given σ - 9.0 is a reasonable planning value, the resulting sample size is n - 52. For each of the following, determine whether the sample size would have been larger or smaller than 52, in each instance assuming that all other specifications remain unchanged from those originally cited.

  a. The β risk has been specified at the .02 level instead of .05 level. _____

  b. The β risk had been specified at the value $\mu = \mu_1$ - 47 instead of the value 45. _____

  c. A larger planning value, say $\sigma = 11.0$, had been used for the population standard deviation. _____

F.8 Consider a two-sided test involving the alternatives H_0: μ - 70 and H_1 : $\mu \neq 70$. The α risk is to be controlled at the .05 level when $\mu = \mu_0$ - 70 and the β risk is to be controlled at the .05 level when $\mu = \mu_1$ - 80. To utilize (11.20) in your Text to compute the necessary sample size, the values z_0 = _____ and z_1 - _____ must first be obtained. Hence, if $\sigma = 25$ is a reasonable planning value, the resulting sample size is n = _____.

F.9 After the sample size has been calculated and the sample outcome obtained, is the planning value for σ or the sample standard deviation, s, used in conducting the appropriate test? _____ Does this affect the β risk at μ_1? _____ Explain. _____
_____√

»Problem 10

a. Recall the situation in Problem 1. Assume that the sample size
   for this upper-tail test has not yet been determined, that $\sigma =$
   3.6 is a reasonable planning value, that the α risk is to be
   controlled at the .10 level when $\mu = \mu_0 = 15.0$, and that the β
   risk is to be controlled at the .005 level when $\mu = \mu_1 = 17.0$.

   i. Find the necessary sample size to satisfy both risk
      specifications. _____ √

   ii. Find the necessary sample size if the β risk specification
       referred to $\mu = \mu_1 = 16.0$ instead of 17.0, everything else
       remaining the same. _____

   iii. From a comparison of the results in (i.) and (ii.) what
        generalization is suggested? _____
        _____ √

b. Recall the situation in Problem 2. Assume that the sample size
   for this lower-tail test has not yet been determined, that $\sigma =$
   4.0 is a reasonable planning value, that the α risk is to be
   controlled at the .01 level when $\mu = \mu_0 = 16$ (as in Problem 2),
   and that the β risk is to be controlled at the .01 level when μ
   $= \mu_1 = 13.0$.

   i. Explain what is meant by the β risk specification in the
      context of this problem. _____

   ii. Find the necessary sample size to satisfy both risk
       specifications. _____

   iii. Suppose the sample size in this situation unwittingly had been
        set at n = 64 (as was done in Problem 2). What can be said
        about the ability to control the α and β risks at the levels
        specified here with a sample size n = 64. _____
        _____ √

c. Recall the situation in Problem 3. Assume that the sample size for this two-sided test has not yet been determined, that $\sigma = 1.0$ is a reasonable planning value, that the α risk is to be controlled at the .10 level when $\mu = \mu_0 = 13.0$, and that the β risk is to be controlled at the .05 level when the mean length of delivered trout differs by one-third inch from that specified in the contract.

   i. At what value $\mu = \mu_1$ is the β risk being controlled here? _____ Explain. _____

_____

  ii. Find the necessary sample size to satisfy both risk specifications. _____

 iii. Find the necessary sample size if the β risk is to be controlled at the .01 level instead of the .05 level, everything else remaining the same. _____

  iv. From a comparison of the results in (ii.) and (iii.), what generalization is suggested? _____

_____ ✓

Answers to Chapter Reviews and Problems

A.1 H_0: $\mu \leq \mu_0$; H_1: $\mu > \mu_0$; H_0: $\mu = \mu_0$; H_1: $\mu \neq \mu_0$

A.2 H_0 denotes some standard against which μ, the mean of the _ _ population, is to be compared.

A.3 null

A.4 The alternative conclusions here are not mutually exclusive. For example, if $\mu = 750$, then both alternatives H_0: $\mu \geq 700$ and H_1: $\mu < 800$ are correct. To be appropriate for statistical tests, the alternative conclusions or hypotheses must be mutually exclusive.

A.5 H_1; H_0; H_0; H_1

A.6 α; β

A.7 H_0: $\mu \geq 12.0$; H_1: $\mu < 12.0$

A.8 H_0: $\mu \leq 10.0$; H_1: $\mu > 10.0$

A.9 H_0: $\mu \leq 30$; H_1: $\mu > 30$

A.10 A Type I error means concluding that $\mu > 30$ bushels per acre when, in fact, $\mu \geq 30$ bushels per acre. A Type II error means concluding that $\mu \leq 30$, when, in fact, $\mu > 30$. No; if $\mu > 30$ the agent could correct conclude that $\mu > 30$ (i.e., select H_1 as correct when, in fact, H_1 is correct and, similarly, if $\mu \leq 30$ he/she could correctly conclude that $\mu \leq 30$. (See Table 12.1 in your Text.)

A.11 H_0: $\mu \geq 200$; H_1: $\mu < 200$

A.12 Type I error; since H_1 was concluded (the alternate supplier was used so it must have been concluded that $\mu < 200$) when, in fact, H_0 is correct (H_0: $\mu \geq 200$ is correct when $\mu = 223$).

A.13 No; since H_1 was concluded when, in fact, H_1 is correct (H_1: $\mu \leq 200$ is correct when $\mu = 185$).

A.14 statistical decision rule

A.15 action limit; A; test statistic

A.16 The acceptance region encompasses the range of values of the test statistic for which alternative 80 is concluded; in this case, the range $X > 22$. The rejection region encompasses the range of values for which H_1 is concluded; in this case, the range $X < 22$.

A.17 In each case here the action limit(s) is (are) set such that alternative H_1 is concluded only if the test statistic X clearly departs from the value of the standard μ_0 in the direction specified in H_1. (See Figure 11.1 in your Text.)

a. The action limit, A, should be some number larger than 1,000.

b. The action limit, A_1, should be some number smaller than 4,500 and the other action limit, A_2, should be some number larger than 4,500.

c. The action limit, A, should be some number smaller than 2,700.

B.1 $\alpha = .10$ when $\mu = \mu_0 = 30$

B.2 a. approximately normal

b. $E(\overline{X}) = \mu_0 = 65.0$

c. $s(\overline{X}) = 15/\sqrt{100} = 1.5$

B.3 standardized; z*

B.4 95th; lf z* ≤ 1.645, conclude $H_0(\mu \le 65.0)$; If z* > 1.645, conclude $H_1(\mu > 65.0)$.

B.5 1.282; -1.282; In both upper-tail and lower-tail tests, it is desired to avoid concluding H_1 unless the sample evidence clearly points to H_1. In an upper-tail test this means limiting the rejection region to the upper .10 tail area of the sampling distribution of z* (or In a lower-tail test this means limiting the rejection region to the_lower .10 tail area of the sampling distribution of z* (or X).

B.6 $\alpha = .02$; From (11.7) in your Text, A = μ_0 + z(1 - α)s($\overline{X}$). Since, in this case, A = 47.054, μ_0 = 45 and s($\overline{X}$) = 12/$\sqrt{144}$ = 1.0, we obtain A = 47.054 = 45 + z(1 - α)(1.0)1. Hence, z(1 - α) = 2.054, which is the position of the 98th percentile of the standard normal distribution. Thus, if 1 - α = .98, then α = .02.

B.7 z(α) = z(.05) = -1.645; Thus, the appropriate decision rule is: If z*> -1.645, conclude $H_0(\mu \ge 200)$; lf z* < -1.645, conclude $H_1(\mu < 200)$.

B.8 s($\overline{X}$) = 50/$\sqrt{400}$ = 2.5; z* = (192 - 200)/2.5 = -3.20; H_1; Since z* = -3.20 < -1.645, conclude $H_1(\mu < 200)$.

B.9 A = 200 + (-1.645)(2.5) = 195.9

B.10 large

B.11 z(.005) = -2.576; z(.995) = 2.576; thus, the decision rule is: If $|z*| \le 2.576$, conclude $H_0(\mu = 12.5)$; if $|z*| > 2.576$, conclude $H_1(\mu \ne 12.5)$.

B.12 s($\overline{X}$) = 10.5/$\sqrt{49}$ = 1.5; z* = (16.2 - 12.5)/1.5 = 2.467; H_0; since $|z*|$ = 2.467 < 2.576, conclude $H_0(\mu = 12.5)$.

Problem 1

   a. H_0: $\mu \leq 15$; H_1: $\mu > 15$

   b. i.

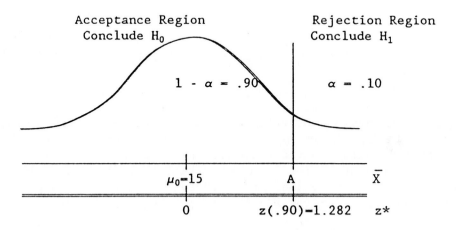

Figure 11.3

    ii. If $z* \leq 1.282$, conclude $H_0(\mu \leq 15)$; If $z* > 1.282$, conclude $H_1(\mu > 15)$.

   c. $s(\overline{X}) = 3.6/\sqrt{36} = .60$; thus, $z* = (\overline{X} - \mu_0)/s(\overline{X}) = (16.4 - 15.0)/.60 = 2.33$.

   d. Since $z* = 2.33/1.282$, conclude H_1 $(\mu > 15)$, the mean percent sugar content in this shipment is greater than 15 percent.

Problem 2

   a. H_0: $\mu \geq 16$; H_1: $\mu < 16$

b. i.

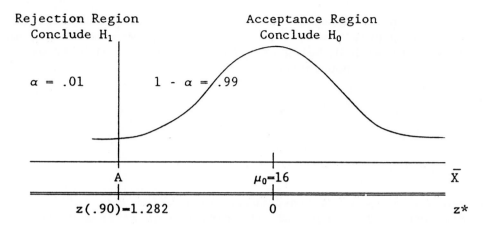

Figure 11.4

ii. If $z* \geq -2.326$, conclude $H_0(\mu \geq 16)$; if $z* < -2.326$, conclude $H_1(\mu < 16)$.

c. $s(\overline{X}) = 4.0/\sqrt{64} = .50$; Thus, $z* = (15.07 - 16.0)/.50 = -1.86$.

d. No; since $z* = -1.86 > -2.326$ we conclude $H_0(\mu \leq 16)$, the mean assembly time with the new procedure is not less than with the old procedure.

Problem 3

a. $H_0: \mu = 13.0$; $H_1: \mu \neq 13.0$

b. $z(1 - \alpha/2) = z(.95) = 1.645$; the decision rule is: if $|z*| \leq 1.645$, conclude $H_0(\mu = 13.0)$; if $|z*| > 1.645$, conclude $H_1(\mu \neq 13.0)$.

c. $s(\overline{X}) = .90/\sqrt{100} = .09$; thus, $z* = (12.91 - 13.0)/.09 = -1.0$.

d. Since $|z*| = 1.0 < 1.645$, conclude $H_0(\mu = 13.0)$, the mean length of delivered trout is 13.0 inches (i.e., the contract requirement is being met).

C.1 The P-value indicates that the probability that $z*$ might have been larger than the observed outcome 1.67, when $\mu = \mu_0 = 45$, is .0475.

C.2 less

C.3 Population 1; since the P-value for population 1 was larger than that for population 2, the sample evidence was relatively more consistent with H_0 being true for population 1 than it was with H_0 being true for population 2.

C.4 H_0

C.5 <; <; less

C.6 .0007; H_1

C.7 Since this is a two-sided test, the P-value must reflect the distance of X from μ_0 (or, the distance of z* from 0) irrespective of the direction of the departure.

C.8 a. $P(Z > 2.33) = .0099$

b. $P(Z < -1.86) = .0314$

c. $2P(Z < -1.00) = 2(.1587) = .3174$

Problem 4

Alternatives: $H_0: \mu \leq 10$; $H_1: \mu > 10$

Decision Rule: If P-value > .05, conclude $H_0(\mu \leq 10)$;
                         If P-value < .05, conclude $H_1(\mu > 10)$

P-value: $s(\overline{X}) = 2.8/\sqrt{49} = .40$; z* = $(9.0 - 10)/.40 = -2.50$;
                   Thus, the P-value = $P(Z > -2.50) = .9938$.

Decision: Since the P-value = .9938 > .05, conclude
                 $H_0(\mu \leq 10)$, the mean weight of parcels shipped on this route does not exceed ten pounds.

Problem 5

Alternatives: $H_0: \mu \le 1.5$; $H_1: \mu > 1.5$

Decision Rule: $t(1 - \alpha;\ n - 1) = t(.90;\ 24) = 1.318$; Thus, the decision rule is:
If $t* \le 1.318$, conclude $H_0(\mu \le 1.5)$;
If $t* > 1.318$, conclude $H_1(\mu > 1.5)$

t*: $s\{\overline{X}\} = 20/\sqrt{25} = .04$;
Thus, $t* = (1.48 - 1.50)/.04 = -.50$.

Decision: Since $t* = -.50 < 1.318$, conclude $H_0(\mu \le 1.5)$, the additive does not yield a mean increase in mileage of more than 1.5 miles per 10 gallons of gasoline used.

Problem 6

Alternatives: $H_0: \mu \ge 30.0$; $H_1: \mu < 30.0$

Decision Rule: $t(\alpha;\ n - 1) = t(.01;\ 8) = -2.896$;
Thus, the decision rule is:
If $t* > -2.896$, conclude $H_0(\mu \ge 30.0)$;
If $t* < -2.896$, conclude $H_1(\mu < 30.0)$

t*: $s\{\overline{X}\} = 4.5/\sqrt{9} = 1.5$;
Thus, $t* = (24.5 - 30.0)/1.5 = -3.667$.

Decision: Since $t* = -3.667 < -2-896$, conclude $H_1(\mu < 30.\overline{0})$, the insecticide is effective in controlling infestation of corn stalks (i.e., the mean number of infested corn stalks per tract is less than 30 when treated with the insecticide).

Problem 7

Alternatives: H_0: $\mu = 60.0$; H_1: $\mu \neq 60.0$

Decision Rule: $t(1 - a/2; n - 1) = t(.975; 15) = 2.131$; Thus,
the decision rule is:
If $|t*| \leq 2.131$, conclude $H_0(\mu = 60.0)$;
If $|t*| > 2.131$, conclude $H_1(\mu \neq 60.0)$

t*: $s(\overline{X}) = 14.5/\sqrt{16} = 3.625$.
Thus, $t* = (71.23 - 60.0)/3.625 = 3.098$.

Decision: Since $|t*| \leq 2.131$, conclude $H_1(\mu \neq 60)$,
the company fleet is not performing at industry
standards.

E.1 If $\overline{X} \leq 215$, conclude H_0 ($\mu \leq 200$); If $\overline{X} > 215$, conclude $H_1(\mu > 220)$.

E.2 Given the decision rule in E.1, $P(H_1; \mu = 220)$ is the probability of
concluding H_1 when the population mean is $\mu = 220$.

E.3 rejection; rejection probability curve

E.4 .2877

E.5 a. Portion in the range where $\mu \leq 200$ (values of μ where H_0 is
correct).

b. Portion in the range where $\mu > 200$ (values of μ where H_1 is
correct). Note that ß risks are given by complements of the
rejection probabilities in the range where $\mu > 200$.

c. About .29; ß risk (Note: From E.4, we know the ß risk at $\mu = 220$
is $P(H_0; \mu = 220) = 1 - P(H_1; \mu = 220) = 1 - .7123 = .2877$.)

d. No; there is no α risk since H_0 is not the correct conclusion if
$\mu = 210$. The risk at $\mu = 210$ is a ß risk.

e. About .05.

E.6 False. Within any given situation, the population mean μ is a fixed value which does not vary. We compute these probabilities for various values of μ because we do not know which one of the many possible values is the actual fixed value of μ in that situation.

E.7 $\sigma(\overline{X}) = 25/\sqrt{100} = 2.5$. Thus, A = 85.0 + 1.645(2.5) = 89.112

E.8 $P(H_1; \mu = 95) = P(Z > (89.112 - 95.0)/2.5) = P(Z > -2.36) = .9909$; the ß risk at $\mu = 95.0$ is 1 - .9909 = .0091.

E.9 $P(H_1; \mu = 84) = P(Z > (89.112 - 84.0)/2.5) = P(Z > -2.04) = .0207$; The ß risk is undefined at $\mu = 84.0$ since H_0 is given by the preceding rejection probability.

E.10 a. The shape of this rejection probability curve dictates the alternatives must have been of the two-sided form. Further, its placement along the μ axis dictates the standard μ_0 must have been 50.0. Hence, the alternatives must have been $H_0: \mu = 50$ and $H_1: \mu \neq 50$.

b. .01; α risk

c. About 1 - .70 = .30; ß risk

E.11 increase

Problem 8

a. $H_0: \mu \geq 26.0$ hours per gallon (adopt the new motor)
   $H_1: \mu < 26.0$ hours per gallon (do not adopt the new motor)

b. A = $\mu_0 + z(\alpha)\sigma(\overline{X}) = 26.0 - 1.645(.4) = 25.34$; thus, the decision rule for the test statistic X is: If X ≥ 25.34, conclude $H_0(\mu \geq 26.0)$; if X < 25.34, conclude $H_1(\mu < 26.0)$.

c. i. and ii.   Rejection Region   Acceptance Region
                Conclude $H_1(\mu<26)$ Conclude $H_0(\mu\geq26)$

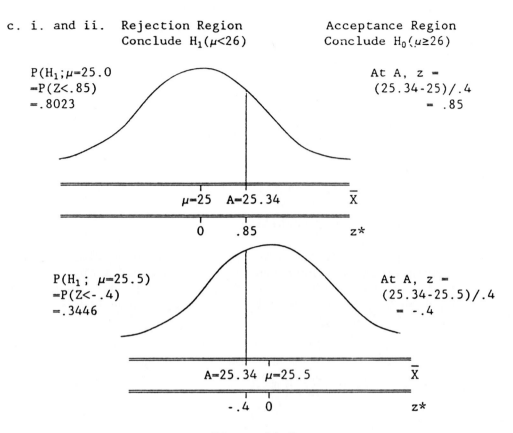

$P(H_1;\mu=25.0$ At A, z =
$=P(Z<.85)$ $(25.34-25)/.4$
$=.8023$ $= .85$

$\mu=25$ A=25.34 $\overline{X}$

0   .85   z*

$P(H_1;\mu=25.5)$ At A, z =
$=P(Z<-.4)$ $(25.34-25.5)/.4$
$=.3446$ $= -.4$

A=25.34 $\mu=25.5$ $\overline{X}$

-.4 0   z*

Figure 11.5

iii. μ $P(H_1;\mu)$ μ $P(H_1;\mu)$

        24.75   .9306   25.75   .1539
        25.00   .8023   26.00   .0500
        25.25   .5871   26.25   .0113
        25.50   .3446   26.00   .0019

d. i., ii. and iii.

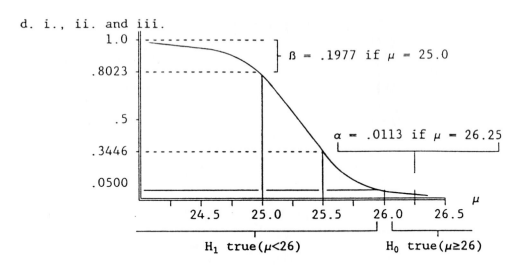

Figure 11.6

iv. The general shape of the rejection probability curve here
reflects that this is a one-sided lower-tail test. Recall
that the rejection probability curve shown in Figure 12.11 in
your Text is based on a one-sided upper-tail test. Notice
that, for this lower-tail test, the lower is the mean hours
of use per gallon, the more probable it is that $H_1 : \mu < 26.0$
will be concluded. Conversely, the higher is the mean hours
of use per gallon, the less probable that $H_1: \mu < 26.0$ will
be concluded. Thus, it is reasonable that rejection
probability curves for one-sided lower-tail tests tend to
fall from left to right.

Problem 9

a. $H_0: \mu = 0$; $H_1: \mu \neq 0$

b. i. A Type I error means to conclude that mean family income has
changed ($\mu \neq 0$) when it has not ($\mu = 0$).

   ii. A Type II error means to conclude that mean family income has
not changed ($\mu = 0$) when it has ($\mu \neq 0$).

c. $A_1 = 0-1.96(1,000/\sqrt{2,500}) = -39.2$; $A_2 = 0+1.96(1,000/\sqrt{2,500}) = 39.2$; thus, the decision rule for the $\bar{X}$ test statistic is:
If $-39.2 \leq \bar{X} \leq 39.2$, conclude H_0 ($\mu = 0$); if $\bar{X} < -39.2$ or $\bar{X} > 39.2$, conclude H_1 ($\mu \neq 0$).

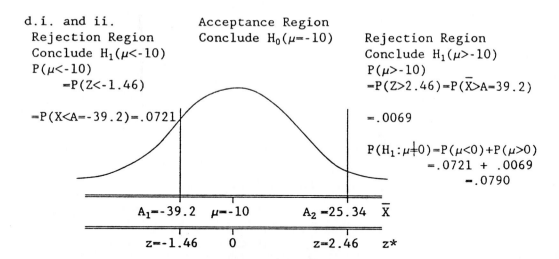

d.i. and ii.
Rejection Region
Conclude $H_1(\mu < -10)$
$P(\mu < -10)$
   $= P(Z < -1.46)$

$= P(X < A = -39.2) = .0721$

Acceptance Region
Conclude $H_0(\mu = -10)$

Rejection Region
Conclude $H_1(\mu > -10)$
$P(\mu > -10)$
$= P(Z > 2.46) = P(\overline{X} > A = 39.2)$

$= .0069$

$P(H_1 : \mu \neq 0) = P(\mu < 0) + P(\mu > 0)$
               $= .0721 + .0069$
               $= .0790$

$A_1 = -39.2 \quad \mu = -10 \qquad A_2 = 25.34 \quad \overline{X}$

$z = -1.46 \quad\quad 0 \qquad\qquad z = 2.46 \quad z*$

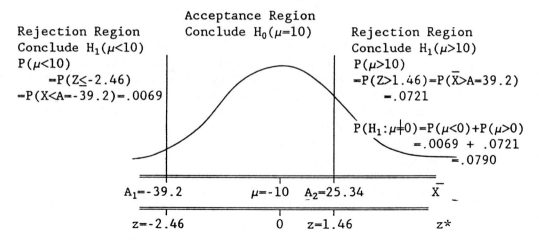

Rejection Region
Conclude $H_1(\mu < 10)$
$P(\mu < 10)$
   $= P(Z \leq -2.46)$
$= P(X < A = -39.2) = .0069$

Acceptance Region
Conclude $H_0(\mu = 10)$

Rejection Region
Conclude $H_1(\mu > 10)$
$P(\mu > 10)$
$= P(Z > 1.46) = P(\overline{X} > A = 39.2)$
   $= .0721$

$P(H_1 : \mu \neq 0) = P(\mu < 0) + P(\mu > 0)$
               $= .0069 + .0721$
               $= .0790$

$A_1 = -39.2 \qquad\qquad \mu = -10 \quad A_2 = 25.34 \qquad \overline{X}$

$z = -2.46 \qquad\qquad 0 \quad z = 1.46 \qquad\qquad z*$

Figure 11.7

iii. With $\mu_0 = 0$, we find $P(H_1 ; \mu = -10) = P(H_1 ; \mu = 10) = .079$. This
suggests that, for two-sided tests, rejection probabilities
for values of μ equidistant above and below μ_0 are equal,
leading to rejection probability curves that are symmetrical
about μ_0.

e.i.

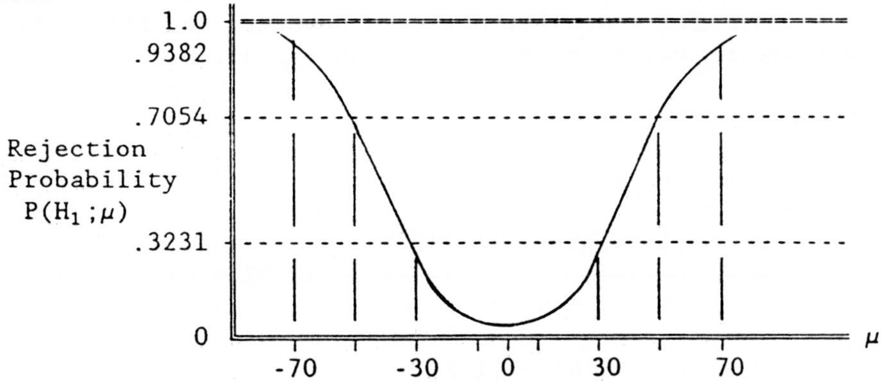

Figure 11.8

   ii. Yes; ß risks are substantial if there has been a slight change in mean income. For example, the ß risk is 1 - .079 = .921 if there has been a change of 10 in mean income in either direction.

  iii. No; ß risks are quite small if there has been a large change in mean income. For example, the ß risk is just 1 - .9382 = .0618 if there has been a change of 70 in mean income in either direction.

F.1 σ, the population standard deviation

F.2 larger; μ_1

F.3 $\alpha = P(H_1; \mu_0 = 7.5) = .025$; $1 - ß = P(H_1; \mu_1 = 8.0) = .95$

F.4 $z_0 = z(1 - \alpha) = z(.975) = 1.96$; $z_1 = z(ß) = z(.05) = -1.645$

F.5 $n = [(.90)^2(|-1.645|+|1.96|)^2]/|8.0-7.5|^2 = 42.11 = 43$. (Note: Throughout the Study Guide, fractional-valued sample sizes are rounded upwards to the nearest integer.)

F.6 1.645; 2.326; 3.971

F.7 a. larger b. larger c. larger

F.8 $z_0 = z(.975) = 1.96$; $z(.05) = -1.645$; $n = 82$

F.9 s; yes, this generally does affect the ß risk because s and the planning value for σ usually differ. However, the effect is slight if the planning value for σ is reasonably good.

Problem 10

a. i. $z_0 = z(1 - \alpha) = z(.90) = 1.282$; $z_1 = z(\text{ß}) = z(.005) = -2.576$;
Thus,
$$n = \frac{(3.6)^2(|-2.576| + |1.282|)^2}{|17.0 - 15.0|^2} = 48.22 = 49 \text{ potatoes}$$

　　ii.
$$n = \frac{(3.6)^2(|-2.576| + |1.282|^2}{|16.0 - 15.0|^2} = 192.90 = 193 \text{ potatoes}$$

　iii. With the levels of α and ß held constant, the required sample size increases as the ß risk is specified at values for the population mean, μ_1, which lie closer to μ_0.

b. i. Management has just a .01 probability of concluding that the mean assembly time with the new procedure is not less than that with the old procedure if, in fact, the mean assembly time with the new procedure is 3 minutes less than that with the old procedure.

　　ii. $z_0 = z(\alpha) = z(.01) = -2.326$; $z_1 = z(1 - \text{ß}) = z(.99) = 2.326$;
Thus,
$$n = \frac{(4.0)^2(|2.326| + |-2.326|^2}{|13.0 - 16.0|^2} = 38.47 = 39 \text{ assembly times}$$

　iii. With a required sample size of n = 39, sampling 64 assembly times provided even more control over the α and ß risks than had been specified. In fact, given these risk specifications, sampling 64 assembly times was uneconomical.

c. i. Since the contract specifies that mean length of delivered trout should equal 13.0 inches, the ß risk is being controlled at the value $\mu_1 = 13.0 + .33 = 13.33$ inches (or, equivalently in terms of determining the required sample size since this is a two-sided test, at the value $\mu_1 = 13.0 - .33 = 12.67$ inches).

ii. $z_0 = z(1 - u/2) = z(.95) = 1.645$; $z_1 = z(\beta) = z(.05) = -1.645$;

$$n = \frac{(1.0)^2(|-1.645| + |1.645|)^2}{|13.30 - 13.0|^2} = 99.39 = 100 \text{ delivered trout}$$

iii.
$$n = \frac{(1.0)^2(|-2.326| + |1.282|)^2}{|13.33 - 13.0|^2} = 144.8 = 145 \text{ delivered trout}$$

iv. With the value of μ, μ_1 and the α level held constant, the required sample size increases as the β risk at μ_1 is more tightly controlled (i.e., as the probability of a Type II error at μ_1 is specified to be smaller in magnitude).

13

INFERENCES FOR
POPULATION PROPORTION

A. POPULATION PROPORTION, SAMPLE STATISTICS, AND THE SAMPLING DISTRIBUTIONS OF f AND $\bar{p}$

(Text: Chapter 13, Sections 13.1 and 13.2)

Review of Basic Concepts

A.1 When interest centers on a population proportion p, one sample statistic with which we are concerned is the number of occurrences in a random sample of f observations having the characteristic reflected by the population proportion. The number of occurrences is denoted by the symbol

A.2 Given a random sample of size n and the number of occurrences f, the sample proportion   p = _____ .

A.3 A forester is studying trees in a very large timber plot. A characteristic of concern is whether or not a tree is diseased. A random sample of 100 trees is tested for the disease. Identify what each of the following denotes in the context of this study.

    a. p _____

    b. f _____

    c. $\bar{p}$ _____

A.4 It is known that 22.5 percent of the applicants for jobs with a company have experience related to the job for which they applied. In a random sample of 75 applicants, it was found that 18 had related experience. In this situation, p = _____ , f = _____ , and p = _____

A.5 When a random sample of size n is selected from an infinite population corresponding to a Bernoulli process with parameter p, the mean of the sampling distribution of f is $E(f)$ = _____ and the mean of the sampling distribution of p is $E(p)$ = _____ .

A.6 For the situation in A.5, the variance of the sampling distribution of f is $\sigma^2(f)$ = _____ and the variance of the sampling distribution of $\bar{p}$ is $\sigma^2(\bar{p})$ = _____.

A.7 For the situation in A.5, the sampling distributions of f and $\bar{p}$ are given by the _____ probability distribution. When the sample size is sufficiently large, these sampling distributions are approximately _____.

A.8 Suppose p= .15 and n=_20, so we have the binomial probability $P(f \leq 2)$ = .4049. Then, $P(p \leq$ _____) = .4049.

A.9 If p = .15 and n = 20, the sampling distribution of $\bar{p}$ is skewed to the _____. Will the skewness of this distribution increase or decrease if p remains unchanged and n is increased to 100? _____

A.10 According to the working rule in your Text, in order to use_the normal approximation to the sampling distribution of f and p, both np and n(1 - p) should be greater than or equal to _____.

A.11 To compute the standardized variable for approximating the f and $\bar{p}$ sampling distributions, we use the formula Z_= _____ for f and the formula Z = _____ for p.

A.12 When a continuous distribution is used to approximate a discrete distribution and the sample size is relatively small, the approximation can by improved by utilizing the _____ for _____ in the approximation. Suppose a random sample of size n = 20 is to be selected from an infinite population where p = .3. In this case, the normal approximation to the binomial probability $P(p \leq .25)$ = $P(f \leq 5)$ can be improved by using the normal probability $P(X \leq$ _____).

A.13 For_the situation in A.4, describe the exact sampling distribution of p, the proportion of applicants in the random sample who have experience related to the job for which they applied.

a. Form: _____

b. Mean: _____

c. Standard Deviation: _____

A.14 For the situation in A.4, can the sampling distribution of $\bar{p}$ be approximated adequately by a normal distribution? _____ Explain. _____ .

A.15 The sampling distribution of $\bar{p}$ when sampling from a finite population can be approximated by a normal distribution provided the sample size is _____ and the population size N is _____ relative to the sample size.

A.16 A mail order firm accepts payment by a certain credit card. The probabilities that a credit card payment contains a fraudulent card number is .002. It is desired to obtain the probability that five or more fraudulent card numbers are contained in a random sample of 1,000 payments. Using the Poisson approximation here, λ = _____. Hence, from Table C-6 in your Text, $P(f \geq 5)$ = _____ . ✓

»Problem 1

a. Ten percent of the items produced by a process are defective. An inspector takes a random sample of n = 4 items and determines the number and proportion defective in the sample.

   i. Obtain the sampling distributions of f and $\bar{p}$. ✓

   ii. Obtain the mean, variance, and standard deviation of each sampling distribution in (i.), using (12.4) in your Text.

$E\{f\}$ = _____ $E\{\bar{p}\}$ = _____

$\sigma^2\{f\}$ = _____ $\sigma^2\{\bar{p}\}$ = _____

$\sigma\{f\}$ = _____ $\sigma\{\bar{p}\}$ = _____

   iii. Find the following probabilities.

$P\{f = 3\}$ = _____ $P\{\bar{p} = 0\}$ = _____

$P\{f \leq 2\}$ = _____ $P\{\bar{p} \leq .5\}$ = _____

$P\{f \geq 1\}$ = _____ $P\{\bar{p} > .5\}$ = _____ ✓

b. In a work sampling study, a machine is observed at random points in time to determine the proportion of time it is operating. Suppose the machine actually operates 80 percent of the time. A sample of 50 observations will be taken.

   i. Is it appropriate to use the normal approximation for the sampling distribution of f in this study? _____ Explain. _____

  ii. Use the normal approximation with a correction for continuity to find the probability that the machine will be found to be operating 45 times or less. _____ √

 iii. Use the normal approximation with a correction for continuity to find the probability that the machine will be found to be operating 35 times or less. _____ √

B. ESTIMATION OF POPULATION PROPORTION
(Text: Chapter 13, Section 13.3)

Review of Basic Concepts

B.1 A study was undertaken to estimate the proportion of adult residents in a TV viewing area with an interest in watching televised soccer. Based on a random sample of N = 800 residents, the following 90 percent confidence interval was obtained: .206 ≤ p ≤ .254. Interpret this confidence interval. _____
_____

B.2 It is desired to construct a 99 percent confidence interval for p, the proportion of a very large labor force currently unemployed, based on a random sample of n = 400 members of the labor force in which f = 50 were unemployed. Given p = 50/400 = .125, the estimated standard deviation s(p) = _____.

B.3 For the situation in B.2, the appropriate confidence limits are
_____ _____.

B.4 In planning a sample size when no reliable planning value for p exists, the value p = _____ can be used to determine a conservatively large sample size. If p is actually equal to .10, will the sample size determined with the preceding planning value likely result in a confidence interval which is considerably wider or narrower than specified? _____ √

»Problem 2

The ACME Company wishes to estimate the proportion of operating time a computer facility shared with another company is used by ACME. A random sample of operating moments is to be selected throughout the year, and the proportion of observations showing ACME use (p) is to be used as an estimate of the population proportion. It is desired to have an estimate of this proportion within ±.02, with a confidence coefficient of 95 percent.

a. In the past, ACME's utilization of the facility has amounted to at least 60 percent of the operating time. Using p = .60 as a planning value, determine the sample size that should be used in this study, rounded to the nearest 100 operating moments. _____ √

b. 2,300 random observations were made, and ACME was found to be using the computer during 1,472 of these. Construct a 95 percent confidence interval for the proportion of operating time that ACME used the computer facility during the year. _____ √

c. Does the preceding confidence interval have the desired precision ± .02 at the 95 percent confidence level? _____ Explain. _____

_____ √

C. STATISTICAL TESTS FOR p

(Text: Chapter 13, Section 13.4)

Review of Basic Concepts

C.1 Suppose that an admissions officer at a university wishes to test
whether the proportion of incoming freshman who receive a score of
600 or more on their entrance exam is .25 or more against the
alternative that the proportion is less than .25. The appropriate
alternatives here are H_0: _____
and H_1: _____

C.2 Consider a statistical test for p involving the alternative H_0: $p \leq$
.4 and H_1: $p > .4$. The α risk here is to be controlled at the .02
level when p – po – .4. A random sample of size n – 400 is to be
selected. The appropriate decision rule, stated in terms of the
standardized test statistic, is: _____
_____ .

C.3 For the situation in C.2, the resulting sample proportion was p –
.44. The standardized test statistic is then $z* = (p- -p)/\sigma(p) =$
$(.44 - .4)/\sigma(p)$, where $\sigma(p) =$_____. Thus, $z* =$_____
and alternative _____ is concluded.

C.4 For the situation in C.2, the rejection probability curve for the
appropriate decision rule tends to _____ from left
to right, reflecting that concluding H_1 becomes progressively
_____ probable for increasingly larger values of p.

C.5 In selecting between the two-sided alternatives H_0: $p = .18$ and H_1:
$P \neq 18$, the α risk is specified at the 5.0 level when $p = p_0 = .18$.
A random sample of size n = 300 is to be selected. The appropriate
decision rule, stated in terms of the standardized test statistic,
is: _____
_____

C.6 For the situation in C.5, 39 of the sample items possessed the
characteristic of interest. Thus, $p =$_____, $|z*| =$
_____ and alternative _____ is concluded.

C.7 For the situation in C.5 and C.6, the appropriate P-value is
_____ .

C.8 For the situation in C.1, suppose it is desired to control the probability of concluding the proportion is less than .25 when it actually is equal to .25 at the .05 level. Also, suppose it is desired to control the probability of concluding the proportion is .25 or more when it actually is equal to .20 at the .02 level. Identify each of the following, all of which are needed to determine the necessary sample size for this one-sided lower-tail test.

a. p_0 = _____   d. p_1 = _____

b. α = _____   e. β = _____

c. z_0 = _____   f. z_1 = _____

C.9 For the situation in C.1 and C.8, the necessary sample size is n = _____. √

»Problem 3

A retail store manager, contemplating changing the respective locations of various departments within the store, has preliminary sketches of the proposed new store layout drawn up. Since such a renovation is costly, the manager intends to carry it out only if the proportion of shoppers who prefer the new layout to the current one exceeds .75. Thus, the manager wishes to conduct a test involving the alternatives H_0: p ≤ .75 and H_1: p > -75. The manager wishes to control the risk of concluding the proportion exceeds .75 at the .05 level when p = p_0 = .75. Also, she wishes to control the risk of concluding the proportion does not exceed .75 at the .10 level when p = p_1 = .85.

a. Determine the required sample size given these risk specifications. _____
   _____ . √

b. i. What is the appropriate decision rule for this test, stated in terms of the standardized test statistic? _____
   _____

   ii. Given the sample 'size in (a.), what is the position of the action limit, A, that forms the boundary between the acceptance region and the rejection region on the p- scale?
   _____ .

    iii. Determine the rejection probability for this decision rule if p = .775. (Note: You will need to draw a figure analogous to Figure 12.6(b) in your Text but which reflects that this is an upper-tail test.) √

    iv. Determine the rejection probability for this decision rule if p = .800.

    v. Determine the rejection probability for this decision rule if p = .825.

    vi. Using the preceding rejection probabilities along with your knowledge of the error specifications that led to this decision rule, sketch the resulting rejection probability curve. √

  c. Suppose that a random sample of n - 137 shoppers is shown preliminary sketches of the proposed new layout, of which 118 preferred the new layout to the current one. What should the manager conclude? _____ Explain. _____

  d. Determine the P-value for this test. _____ √

Answers to Chapter Reviews and Problems

A.1 f

A.2 $\bar{p} = f/n$

A.3 a. p denotes the proportion of trees in the plot with the disease.

  b. f denotes the number of trees in the sample of 100 with the disease.

  c. $\bar{p}$ denotes the proportion of trees in the sample of 100 with the disease.

  (Note: In the context of this example, it would also have been acceptable to define the parameter p and the sample statistics f and p in terms of trees in the plot without the disease.)

A.4 p=.225; f = 18; $\bar{p}$= 18/75 -.24

A.5 $E\{f\} = np;\ E\{\bar{p}\} = p$

A.6 $\sigma^2\{f\} = np(1 - p);\ \sigma^2\{\bar{p}\} = p(1-p)/n$

A.7 binomial; normal

A.8 $P(\bar{p} \le 2/20 = .10) = .4049$

A.9 right; decrease

A.10 5

A.11 $(f - np)\ /\sigma\{f\}$, where $\sigma\{f\} = \sqrt{np(1 - p)}$; $(\bar{p} - p)\ /\sigma\{\bar{p}\}$, where $\sigma\{\bar{p}\}$ $= \sqrt{p(1 - p)/\ n}$

A.12 correction for continuity; $P(X \le 5.5)$

A.13 a. binomial; b. .225; c. .0482

A.14 Yes; since both $np = 75(.225) = 16.875$ and $n(1 - p) = 75(.775) =$ 58.125 exceed 5.

A.15 large; large

A.16 $\lambda = 1,\ 000(.002) = 2.0;\ P(f \ge 5) = .0526$

Problem 1

a. i. $n = 4$; $p = .10$; Using Table C-5, we obtain:

Sampling Distribution of f		Sampling Distribution of p	
f	$P(f)$	$\bar{P}$	$P(\bar{p})$
0	.6561	0	.6561
1	.2916	.25	.2916
2	.0486	.50	.0486
3	.0036	.75	.0036
4	.0001	1	.0001
Total	1.000	Total	1.000

ii. $E\{f\} = 4(.1) = .4$;
   $\sigma^2\{f\} = 4(.1)(.9) = .36$;
   $\sigma\{\underline{x}\} = \sqrt{.36} = .60$;
   $E\{\underline{p}\} = .1$;
   $\sigma^2\{p\} = (.1)(.9)/4 = .0225$;
   $\sigma\{\overline{p}\} = \sqrt{.0225} = .15$

iii. $P(f = 3) = .0036$;
   $P(f \leq 2) = .6561 + .29161 + .0486 = .9963$;
   $P(\underline{f} \geq 1) = .2916 + .0486 + .0036 + .0001 = .3439$;
   $P(\underline{p} = 0) = .6561$;
   $P(\underline{p} \leq 5) = .6561 + .2916 + .0486 = .9963$;
   $P(p > .5) = .0036 + .0001 = .0037$

b. i. Yes; since $np = 50(.8) = 40 > 5$ and $n(1-p) = 50(.2) = 10 > 5$.

ii. $E\{f\} = 50(.8) = 40$;
   $\sigma\{f\} = \sqrt{50(.8)(.2)} = 2.83$;
   Correcting for continuity gives X = 45.5 in the normal approximation. Thus, $z = (45.5-40)/2.83 = 1.94$. Hence, $P(X \leq 45.5) = P(Z \leq 1.94) = .9738$.

iii. Correcting for continuity gives X = 35.5 in the normal approximation. Thus, $z = (35.5-40)/2.83 = -1.59$. Hence, $P(X \leq 35.5) = P(Z \leq -1.59) = 0559$.

B.1 With 90 percent confidence, we can state that the percent of adult residents in the viewing area with an interest in watching - televised soccer is between 20.6 percent and 25.4 percent. The confidence coefficient indicates that if this estimation procedure were repeated with many independent random samples of the same type used for this study, the resulting interval estimates would be correct (i.e., would contain p) in 90 percent of the cases.

B.2 $s\{\overline{p}\} + \sqrt{.125(1 - .125)/(400-1)} = .0166$

B.3 $.125 \pm 2.576(.0166)$

B.4 .5 ; narrower

Problem 2

    a. $z(.975) = 1.96$; $p = .6$; $h = .02$; $n = (1.96)^2(.6)(1-.6)/(.02)^2 = 2,304$ $\approx 2,300$ operating moments.

    b. $\bar{p} = 1,472/2,300 = .64$; $s(\bar{p}) = \sqrt{.64(1-.64)}/(2300-1) = .0100$; The confidence limits are $.64 \pm 1.96(.0100)$; Thus, the 95 percent confidence interval for p is $.6204 \le p \le .6596$.

    c. Yes; the half-width .0196 is slightly better than the desired precision of $\pm$ .02.

C.1 H_0: $p \ge .25$; H_1: $p < .25$

C.2 If $z^* \le 2.054$, conclude $H_0(p \le .4)$; lf $z^* > 2.054$, conclude $H_1(p > .4)$.

C.3 $\sigma(\bar{p}) = \sqrt{p_0(1 - p_0)/n} = \sqrt{.4(1 - .4)/400} = .0245$; $z^* = (\bar{p} - p_0)/\sigma(\bar{p}) = (.44-.40)/.0245 = 1.633$; H_0

C.4 rise; more

C.5 If $\left| z^* \right| \le 1.96$, conclude $H_0(p = .18)$; If $\left| z^* \right| > 1.96$, conclude $H_0(p \ne .18)$

C.6 $\bar{p} = 39/300 = .13$; $\left| z^* \right| = 2.252$ (Note: $\sigma(\bar{p}) = \sqrt{.18(1-.18)}/300 = .0222$ and, hence, $z^* = (.13 - .18)/.0222 = -2.252$); H_1

C.7 $(2)P(Z > 2.25) = 2(.0122) = .0244$

C.8 a. $p_0 = .25$

    b. $\alpha = .05$

    c. $z_0 = z(.05) = -1.645$

    d. $p_1 = .20$

    e. $\beta = .02$

    f. $z_1 = z(.98) = 2.054$

C.9 $n = [|2.054|\sqrt{.20(1-.20)} + |-1.645|\sqrt{.25(1-.25)}]^2/|.20-.25|^2 = 941.2$
= 942 incoming freshmen

Problem 3

a. $z_0 = z(.95) = 1.645$; $z_1 = z(.10) = -1.282$; Thus, n =

$[|-1.282|\sqrt{.85(1-.85)} + |1.645|\sqrt{.75(1-.75)}]^2/|.85-.75|^2 = 136.9$
= 137 shoppers

b. i. If $z* \leq 1.645$, conclude H_0 (p $\leq$ -75);
If $z* > 1.645$, conclude H_1 (p > .75).

ii. $\sigma(\bar{p}) = \sqrt{.75(1 - .75)/137} = .0370$; Thus, A = $p_0 + z(1-\alpha)\sigma(p) = .75 + 1.645(.0370) = .8109$.

iii., iv. and v.

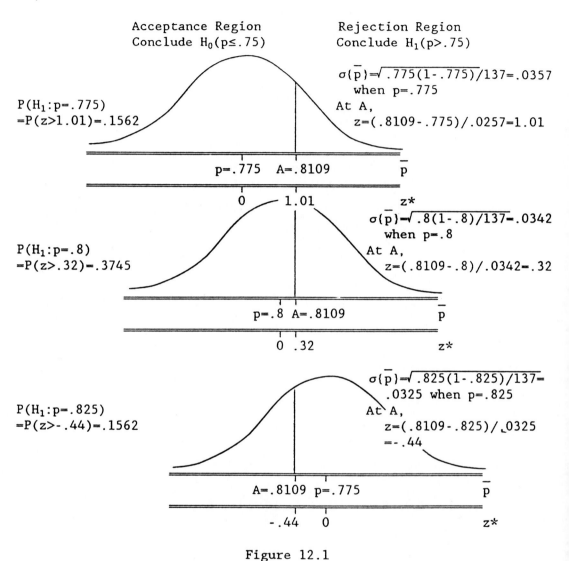

Acceptance Region
Conclude $H_0(p\leq.75)$

Rejection Region
Conclude $H_1(p>.75)$

$\sigma(\bar{p})=\sqrt{.775(1-.775)}/137=.0357$
when $p=.775$
At A,
$z=(.8109-.775)/.0257=1.01$

$P(H_1:p=.775)$
$=P(z>1.01)=.1562$

p=.775   A=.8109 $\bar{p}$

0   1.01   z*

$\sigma(\bar{p})=\sqrt{.8(1-.8)}/137=.0342$
when $p=.8$
At A,
$z=(.8109-.8)/.0342=.32$

$P(H_1:p=.8)$
$=P(z>.32)=.3745$

p=.8 A=.8109 $\bar{p}$

0 .32   z*

$\sigma(\bar{p})=\sqrt{.825(1-.825)}/137=$
.0325 when $p=.825$
At A,
$z=(.8109-.825)/.0325$
$=-.44$

$P(H_1:p=.825)$
$=P(z>-.44)=.1562$

A=.8109 p=.775 $\bar{p}$

-.44   0   z*

Figure 12.1

vi.

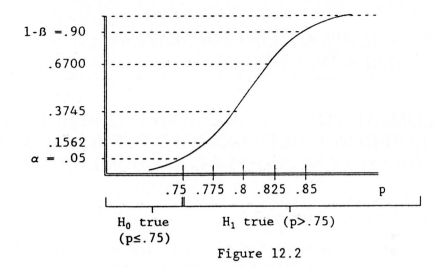

Figure 12.2

c. p = 118/137 = .8613; $\sigma(\bar{p})$ = .0370 when p = p_0 = .75; z* = (.8613-.75)/.0370 = 3.01; Thus, since z* = 3.01>1.645, conclude H_1(p>.75), the proportion of shoppers who prefer the new layout to the current one exceeds .75.

d. The ·P-value equals P(Z>3.01) = 1 - .9987 = .0013

14

COMPARISONS OF TWO POPULATIONS AND OTHER INFERENCES

A. COMPARATIVE STUDIES AND INFERENCES ABOUT DIFFERENCE BETWEEN TWO POPULATION MEANS INDEPENDENT SAMPLES, RANDOMIZED DESIGN

(Text: Chapter 14, Sections 14.1 and 14.2)

Review of Basic Concepts

A.1 If "before" and "after" observations are positively related, the _____ sample method will generally lead to smaller sampling errors than the _____ samples method.

A.2 Indicate whether each of the following studies involves independent or matched samples.

    a. IQ scores of a random sample of 50 students are measured before they enter high school and again after they graduate. The change in each student's IQ score is determined and then the mean change for the sample of 50 students is computed.

    b. IQ scores of a random sample of 140 high school students in a Western city are measured and the mean IQ score is computed. This mean is then compared to the mean IQ score of a separate and unrelated sample of 200 high school students in an Eastern city. _____

    c. The age difference between partners in a marriage is being studied. A random sample of 70 couples is selected and, for each couple, the difference in age between the husband and wife is determined. The mean age difference for all couples in the sample is then computed.

A.3 Suppose a 90 percent confidence interval for the difference $\mu_2 - \mu_1$ is $-308.50 < \mu_2 - \mu_1 < -120.00$, where μ_2 and μ_1 represent the mean per capita income last year in St. Cloud, Minnesota and Duluth, Minnesota, respectively. Interpret this confidence interval.

A.4 By convention in your Text, the mean of population 1 is denoted
_____ mean of population 2 is denoted by _____.
By similar convention, observations from population 1 are denoted
by _____ and observations from population 2 are denoted by
_____. Hence, sample means from populations 1 and 2 are
denoted by _____ and _____, respectively.

A.5 Suppose independent random samples from two normal populations with
the same variance σ^2 are selected and the following sample results
measured.

$$\text{Population 1: } n_1 = 15 \quad \overline{X}_1 = 20.0 \quad s^2_1 = 2.5$$

$$\text{Population 2: } n_2 = 12 \quad \overline{X}_2 = 25.0 \quad s^2_2 = 3.0$$

It is desired to conduct a statistical test involving $\mu_2 - \mu_1$,   the
difference between the means of these two populations, where the
alternatives are H_0: $\mu_2 - \mu_1 = 0$ and H_1: $\mu_2 - \mu_1 \neq 0$.

a. The point estimator of the difference between the means of these
   two populations $(\mu_2 - \mu_1)$ is denoted by _____.   In
   this situation, the point estimate of $\mu_2 - \mu_1$ is _____.

b. In order to conduct the test involving the difference between
   the means of these two populations, we must obtain an estimate
   of their common variance σ^2.   The best unbiased point estimator
   of σ^2, is denoted by _____ and is called the
   _____ or estimator of σ^2.   In this situation, the
   best unbiased point estimate of σ^2 is _____.

c. Once we have computed s^2 we can estimate the variance of the
   sampling distribution of $\overline{X}_2 - \overline{X}_1$, $\sigma^2 (\overline{X}_2 - \overline{X}_1)$, with the
   unbiased estimation of $s^2(\overline{X}_2 - \overline{X}_1)$_____.   In this
   situation, $s^2 (\overline{X}_2 - \overline{X}_1) = $ _____.

d. The appropriate statistical test can then be conducted, since
   the standardized test $(\overline{X}_2 - \overline{X}_1) - 0 / s(\overline{X}_2 - \overline{X}_1)$ is a (n)
   _____ random variable with _____ degrees
   of freedom.

A.6 Suppose large independent random samples are selected from two infinite populations and the following sample results measured.

$$\text{Population 1: } n_1 = 100 \quad \bar{X}_1 = 52 \quad s^2_1 = 125$$

$$\text{Population 2: } n_2 = 400 \quad \bar{X}_2 = 60 \quad s^2_2 = 300$$

Determine each of the following, assuming that the two populations are not normal and do not have equal variances.

a. $\bar{X}_2 - \bar{X}_1 = $ _____

d. $s^2 (\bar{X}_2 - \bar{X}_1) = $ _____

b. $s^2 (\bar{X}_2) = $ _____

e. $s(\bar{X}_2 - \bar{X}_1) = $ _____

c. $s^2 (\bar{X}_1) = $ _____

A.7 Under the conditions set out in A.6, $(\bar{X}_2 - \bar{X}_1) - 0 / s(\bar{X}_2 - \bar{X}_1)$ is approximately a(n)_____ _____ random variable.

A.8 Determine the appropriate value from either the t or Z distribution in each of the following cases. All involve independent samples.

a. A 90 percent confidence interval is to be constructed for the difference between the means of two normal populations with equal variance, where $n_1 = 7$ and $n_2 = 10$. _____

b. A one-sided lower-tail test with $\alpha = .01$ is to be conducted for the difference between the means of two normal populations with equal variance, where $n_1 = 7$ and $n_2 = 10$. _____

c. A two-sided test with $\alpha = .10$ is to be conducted for the difference between the means of two non-normal but symmetrical populations, where $n_1 = 60$ and $n_2 = 90$. _____√

»Problem 1

a. A razor blade manufacturer is studying two types of blades (A,B) to see if the mean number of comfortable shaves obtained from a package of seven blades of Type A (μ_1) is the same as the mean from a package of seven blades of Type B (μ_2). Twenty men who use a safety razor regularly are selected and divided randomly into two groups of ten, one group for each type of blade. Each man is asked to report on the number of comfortable shaves obtained with a package. The reports of the men are summarized as follows:

$$\text{Blade A: } n_1 = 10 \quad \bar{X}_1 = 46.9 \quad s^2 = 196$$

$$\text{Blade B: } n_2 = 10 \quad \bar{X}_2 = 75.6 \quad s^2 = 223$$

Assuming that the number of shaves obtained from a package is approximately normally distributed and the population variances are equal, construct a 99 percent confidence interval for $\mu_2 - \mu_1$. _____ ✓

b. The manager of a theatre chain wishes to compare the mean number of persons per showing among those attending movies rated P or PG (μ_1) to the mean number of persons per showing among those attending movies rated R (μ_2). Independent random samples of showings from each population were selected from last year's records, and the following sample results were obtained.

$$\text{Movies Rated G or PG: } n_1 = 49 \quad \bar{X}_1 = 118 \quad s^2_1 = 637$$

$$\text{Movies Rated R: } n_2 = 64 \quad \bar{X}_2 = 163 \quad s^2_2 = 3,840$$

Construct a 95 percent confidence interval for $\mu_2 - \mu_1$.

_____ ✓

»Problem 2

A trucking firm wishes to test whether the mean service life of the batteries now used in all company trucks (μ_1) is equal to the mean service life of a new battery (μ_2) being promoted by the industry. An experiment was conducted in which independent random samples of each type of battery were measured for service life. The experiment resulted in the following data:

Current Battery: $n_1 = 121$ $\bar{X}_1 = 18.40$ months $s^2_1 = 10.89$

New Battery: $n_2 = 36$ $\bar{X}_2 = 21.30$ months $s^2_2 = 17.64$

Conduct the appropriate statistical test with $\alpha = .02$ when $\mu_2 - \mu_1 = 0$.

Specify the alternatives and decision rule, compute the value of the standardized test statistic, and state the resulting decision.✓

B.   INFERENCES ABOUT DIFFERENCE BETWEEN TWO POPULATION MEANS-MATCHED SAMPLES
(Text: Chapter 14, Section 14.3)

Review of Basic Concepts

B.1 A study was designed to measure the effect of a new diet on weight loss among overweight children. Fourteen randomly selected subjects were weighed before being put on the diet (X_1). The subjects were again weighed after they had been on the diet for a period of sixty days (X_1). In the context of the study, explain what each of the following denotes.
a. D_i _____
b. D _____
c. s^2_D _____
d. $s^2(D)$ _____

B.2 For the situation in B.1, are the matched pairs of observations (X_1, X_2) likely to be positively related? _____
Explain. _____
_____

B.3 For the situation in B.1, $(\bar{D} - \mu_D)/s(\bar{D})$ follows approximately a(n) _____ distribution with _____ degrees of freedom provided the population of differences not differ markedly from a normal distribution.

B.4 For almost all populations of differences, when the sample of differences in a matched sample is reasonably large, $(\bar{D} - \mu_D)/s(\bar{D})$ follows approximately a(n) _____ distribution. ✓

»Problem 3

Twenty-four students in an MBA program were randomly selected to participate in a pedagogical experiment. Students were paired by background and graduate entrance exam scores. One student from each pair was randomly selected and assigned to a business policy course taught by the case method, while the other student was assigned to a class that analyzed an actual business. At the end of the term, the students were given a common exam on business policy. Exam scores for students taught by the case method (X_1) and for those assigned to an actual business (X_2) are presented in Table 14.1.

Table 14. 1

EXAM SCORES FOR 24 STUDENTS IN BUSINESS POLICY COURSE
(See Problem 3)

Pair	(1) Case Method (X_1)	(2) Actual Business (X_2)	(3) D	(4) $\overline{D}-D$	(5) $(D-\overline{D})^2$
1	81	93			
2	85	91			
3	72	68			
4	65	68			
5	99	100			
6	95	92			
7	85	96			
8	84	84			
9	70	79			
10	94	96			
11	90	97			
12	88	80			
		Total			

a. Complete columns 3, <u>4</u>, and 5 in Table 13.1 and compute $\overline{D}$ and s^2_D for this data set. $D =$ _____ ; $s^2_D =$ _____ √

b. Construct a 99 percent confidence interval for the difference in mean scores for the two teaching methods (μ_2 - μ_1), assuming that the population of differences is normal. _____

c. Interpret the confidence interval obtained in part (b.). _____ √

»Problem 4

A random sample of 50 consumers is being used to measure the impact of a new product formulation on consumer attitudes toward a packaged cake mix. Each consumer is asked to rate the taste of the old formulation on a 100 point scale (X_1). Each person is then asked to rate the taste of the new formulation on the same scale (X_2). Based on the 50 sample differences D - X_2 - X_1, the sample results are D - 20.70 and s_D - 10.82. Conduct a test to determine if the new formulation leads to an improved (higher) mean consumer taste rating. Set α = .10 when μ_D = μ_2 - μ_1 = 0. Specify the alternatives and decision rule, compute the value of the standardized test statistic, and state the resulting decision. (Note: The nature of the population of differences is not specified here. However, since the sample size is reasonably large (i.e., n = 50) the test illustrated in Figure 11.2 in your Text is applicable.) √

C. INFERENCES ABOUT DIFFERENCE BETWEEN TWO POPULATION PROPORTIONS
(Text: Chapter 14, Section 14.4)

Review of Basic Concepts

C.1 The procedures discussed in Section 14.4 in your Text for making inferences about the difference between two population proportions require that the two random samples be _____ and _____.

C.2 When large independent random samples are selected from two populations, the variance of the sampling distribution of p_2 - p_1 is the _____ of the variances of the sampling distributions of _____ and _____. Symbolically, $\sigma^2(p_2 - p_1)$ = _____.

C.3 Suppose independent random samples of size $n_1 = 300$ and $n_2 = 500$ have been selected from two infinite populations with resulting sample proportions $p_1 = .15$ and $p_2 = .12$. To construct a 95 percent confidence interval for $p_2 - p_1$ we need an estimator of $\sigma^2(p_2 - p_1)$. In general, this estimator is denoted by _____ and in this example, the estimate is _____. Hence, $s(p_2 - p_1) =$ _____ and the appropriate confidence limits are _____ $\pm$ _____ .

C.4 In selecting between the alternatives H_0: $p_2 - p_1 = 0$ and H_1: $p_2 - p_1 \neq 0$ the value of the test statistic z^* depends, in part, on the magnitude of $s^2(p_2 - p_1)$. Explain what $s^2(p_2 - p_1)$ denotes in the context of this test.

C.5 For the situation in C.4, one component in the formula for $s^2(\overline{p_2}\,\overline{p_1})$ is $p.'$ Explain what p' denotes and how it is computed. _____

C.6 For the situation in C.4, suppose $n_1 = 100$, $n_2 = 200$ and the sample results are $p_1 = .66$ and $p_2 = .60$. Thus, p' _____ $s^2(p_2 - p_1) =$ _____ and $s(p_2 - p_1) =$ _____ when $p_2 = p_1$. $\checkmark$

»Problem 5

A study was conducted to estimate the difference in the proportion of firms which showed an operating loss last year for corporations (p_1) and sole proprietorships (p_2). Independent random samples of 100 firms from each population were selected. Twenty of the sole proprietorships and 10 of the corporations sampled showed an operating loss last year. Construct a 95 percent confidence interval for $p_2 - p_1$. _____ $\checkmark$

»**Problem 6**

In a production run of 6,200 units, 5,000 units were sold to customers under one type of warranty contract and the remaining 1,200 units were sold to customers under a second type of warranty contract. Assuming that production conditions were such that these can be viewed as two independent random samples of size n_1 = 5,000 and n_2 = 1,200, the manufacturer wishes to test whether or not the proportion of warranty adjustments under the first contract (p_1) and the second contract (p_2) are equal. Sales under the first contract resulted in 545 adjustments and sales under the second contract resulted in 168 adjustments. Conduct the appropriate test involving p_2 - pl with α = .02. Specify the alternatives and decision rule, compute z*, and state the resulting decision. √

Answers to Chapter Reviews and Problems

A.1 matched; independent

A.2 a. matched

   b. independent

   c. matched

A.3 With 90 percent confidence, we can state that the mean per capita income in St. Cloud, Minnesota, last year was between $120.00 and $308.50 less than the mean per capita income last year in Duluth, Minnesota.

A.4 μ_1; μ_2 ; X_1; X_2; $\overline{X}_1$; $\overline{X}_2$

A.5 a. $\overline{X}_2 - \overline{X}_1$ = 25.0 - 20.0 = 5.0

   b. $s^2{}_c$; pooled or combined; $s^2{}_c = \dfrac{(15 - 1)(2.5) + (12 - 1)(3.0)}{(15 - 1) + (12 - 1)} =$ 2.72

   c. $s^2(\overline{X}_2 - \overline{X}_1) = s^2{}_c(1-/n_2 + 1/n_1) = 2.72(1/12 + 1/15) = .408$

   d. t; (15 + 12 - 2) = 25

A.6 a. $\overline{X}_2 - \overline{X}_1$ = 60 - 52 = 8

b. $s^2\{\overline{X}_2\}$ = 300/400 = .75

c. $s^2\{\overline{X}_1\}$ = 125/100 = 1.25

d. $s^2\{\overline{X}_2 - \overline{X}_1\}$ = .75 + 1.25 = 2.00

e. $s\{\overline{X}_2 - \overline{X}_1\}$ = $\sqrt{2.00}$ = 1.41 - 4

A.7 standard normal

A.8 a. t(.95; 10 + 7 - 2) = 1.753

b. t(.01; 10 + 7 - 2) = -2.602

c. z(.95) = 1.645

Problem 1

a. $\overline{X}_2 - \overline{X}_1$ = 75.6 - 46.9 = 28.7; t(.995; 10 + 10 - 2) =2.878;
$S^2{}_c\_$= (1$\underline{0}$ - 1) 196 + (10 - 1) 223/ (10 - 1) $\pm$ (10$\_$-1) = 209.5
$s^2\{\overline{X}_2 - \overline{X}_1\}$ = 209.5 (1/10 + 1/10) = 41.9; $s\{\overline{X}_2 - \overline{X}_1\}$ = 6.473.
The confidence limits are 28.7 $\pm$ 2.878(6.473). Thus, the 99
percent confidence interval for $\mu_2 - \mu_1$ is
10.07 $\leq \mu_2 - \mu_1 \leq$ 61.746.

b. $\overline{X}_2 - \overline{X}_1$ = 163 - 118 = 45; z(.975) = 1.9$\underline{6}$0; $s^2\{\overline{X}_2 - \overline{X}_1\}$ = $s^2\{\overline{X}_2\}$
+ $s^2\{X_1\}$ = 3,840/64 + 637/49 = 73.0; $s\{\overline{X}_2 - \overline{X}_1\}$ = $\sqrt{73.0}$ = 8.544.
The confidence limits are 45 $\pm$ 1.960(8.544). Thus, the 95
percent confidence interval for $\mu_2 - \mu_1$ is
28.254 $\leq \mu_2 - \mu_1 \leq$ 61.746.

Problem 2

Alternatives: H_0: $\mu_2 - \mu_1 = 0$
H_1: $\mu_2 - \mu_1 \neq 0$

Decision Rule: $z(1 - \alpha/2) = z(.99) = 2.326$; thus, the decision rule is:

If $|z*| \leq 2.326$, conclude H_0;
If $|z*| > 2.326$, conclude H_1

$z*$: $(\overline{X}_2 - \overline{X}_1) = 21.30 - 18.40 = 2.90$;

$s^2(\overline{X}_2 - \overline{X}_1) = 17.64/36 + 10.89/121 = .580$;

$s(\overline{X}_2 - \overline{X}_1) = \sqrt{.580} = .7616$

Thus, $z* = (2.90 - 0)/.7616 = 3.808$

Decision: Since $|z*| = 3.808 > 2.326$, conclude H_1, the mean service lives of the two batteries are not equal.

B.1 a. D represents the difference, weight after the diet (X_1) minus weight before the diet (X_1) for the ith subject. Hence $D = X_2$ X_1 represents the change in weight for the ith subject.

b. $\overline{D}$ represents the mean of the differences $\underline{D}_i$, weight after the diet minus weight before the diet. Thus, D represents the mean change in weight for the 14 subjects in the sample.

c. $s^2{}_D$ represents the variance of the changes in weight (D's) obtained in the sample of n = 14 subjects.

d. $s^2(\overline{D})$ represents the estimated variance of the sampling distribution of D.

B.2 Yes; unless the effectiveness of the diet varies a great deal across subjects, the subjects who weighed most before the diet are likely to still weigh most after the diet. Similarly, those who weighed least before the diet are likely to still weigh least after the diet.

B.3 t; 14- 1 = 13

B.4 normal

Problem 3

a.

Pair	(3) D	(4) $D-\bar{D}$	(5) $(D-\bar{D})^2$	Pair	(3) D	(4) $D-\bar{D}$	(5) $(D-\bar{D})^2$
1	12	9	81	7	11	8	64
2	6	3	9	8	0	-3	9
3	-4	-7	49	9	9	6	36
4	3	0	0	10	2	-1	1
5	1	-2	4	11	7	4	16
6	-3	-6	36	12	-8	11	121
				Total	36		426

$\bar{D}$ = 36/12 = 3.0; s_D^2 = 426/11 = 38.727

b. $\bar{D}$ = 3.0; t(.995; 11) = 3.106; $s^2(\bar{D})$ = 38.727/12 = 3.227; $s(\bar{D})$ = $\sqrt{3.227}$ = 1.796. The confidence limits are 3.0±3.106(1.796). Thus, the 99 percent confidence interval for $\mu_2 - \mu_1$ is -2.58 ≤ $\mu_2-\mu_1$ ≤ 8.58.

c. With 99 percent confidence, we can state that the mean score of students in the "actual business" course is between 2.58 points less and 8.58 points more than the mean score of students in the "case method" course.

Problem 4

Alternatives: H_0: $\mu_2-\mu_1$ ≤ 0; H_1: $\mu_2-\mu_1$ > 0

Decision Rule: z(1 - α) = z(.90) = 1.282; thus the decision rule is: if z* < 1.282, conclude $H_0(\mu_2-\mu_1$ ≤ 0); if z* > 1.282, conclude $H_1(\mu_2-\mu_1$ > 0)

z*: $\bar{D}$ = 20.70;

$S^2(\bar{D})$ = $(10.82)^2/50$ = 2.341;

$s(\bar{D})$ = $\sqrt{2.341}$ = 1.530;

Thus, z* = (20.70-0)/1.53 = 13.53.

Decision: Since $z* = 13.53 > 1.282$, conclude $H_1(\mu_2 - \mu_1 > 0)$, the mean rating of the new formulation is greater than the mean rating for the old formulation.

C.1 large; independent

C.2 sum; $\bar{p}_2$ and $\bar{p}_1$; $\sigma^2\{\bar{p}_2 - \bar{p}_1\} = \sigma^2\{\bar{p}_2\} + \sigma^2\{\bar{p}_1\}$

C.3 $s^2\{\bar{p}_2 - \bar{p}_1\} = .12(1-.12)/(500-1) + .15(1-.15)/(300-1) = .000638$; $s\{\bar{p}_2 - \bar{p}_1\} = -0253$; $-.03 \pm 1.96(.0253)$

C.4 As illustrated in Figure 14.2 in your Text, determining the value of $z*$ for this test requires knowledge of the sampling distribution of p_2-p_1 when H_0: $p_2-p_1 = 0$ holds._ This distribution is centered at $p_2-p_1 = 0$ with variance $\sigma^2\{p_2-p_1\}$. For this test, $s^2\{p_2-p_1\}$ then denotes the sample estimator of $\sigma^2\{p_2-p_1\}$ then H_0: $p_2-p_1 = 0$ holds.

C.5 For the case when H_0: $p_2-p_1 = 0$ holds, p_1 and p_2 are equal. $\bar{p}'$ is a sample estimator of their common value $p_1 = p_2 = p$. It is called the pooled estimator of_p and is_computed as a weighted average of p_1 and p_2. Thus, $p' = (n_1 p_1 + n_2 p_2)/(n_1 + n_2)$.

C.6 $\bar{p}' = [100(.66)+200(.60)]/(100+200) = .62$; $s^2\{p_2-p_1\} = .62(1 - .62)(1/200 + 1/100) = "003534$; $s^2\{p_2-p_1\} = .0594$

Problem 5

$\bar{p}_2 - \bar{p}_1 = .2-.1 = .1$; $z(.975) = 1.960$;

$s\{\bar{p}_2 - \bar{p}_1\} = \sqrt{[.2(1-.2)]/(100-1) + [.1(1-.1)]/(100-1)} = .0503$.
The confidence limits are $.1 \pm 1.960(.0503)$. Thus, the 95 percent confidence interval for p_2-p_1 is $.0014 \le p_2-p_1 \le .1986$.

Problem 6

Alternatives: H_0: $p_2-p_1 = 0$; H_1: $p_2-p_1 \neq 0$

Decision Rule: $z(1 - \alpha/2) = z(.99) = 2.326$; Thus, the
decision rule is :
If $\left| p^* \right| \leq 2.326$, conclude $H_0(p_2-p_1 = 0)$;
If $\left| p^* \right| > 2.326$, conclude $H_1(p_2-p_1 \neq 0)$;

Z*: $\overline{p}_1 = 545/5000 = .109$; $\overline{p}_2 = 168/1200 = .140$;
Thus, $p_2-p_1 = .140 - .109 = .031$. When $p_2-p_1 = 0$, $p' = [5000(.109)+1200(.140)]/(1/1200+1/5000)$

=.000105, and $s(\overline{p}_2-\overline{p}_1) = \sqrt{.000105} = .0102$.
Thus, $z^* = (1.031-0)/.0102 = 3.039$.

Decision: Since $\left| z^* \right| = 3.039 > 2.326$, conclude
$H_1(p_2-p_1 \neq 0)$, the proportion of warranty
adjustments under the two contracts are not
equal.

15

NONPARAMETRIC PROCEDURES

A. Inferences About η
(Text: Chapter 15, Section 15.1)

Review of Basic Concepts

A.1 Estimation and tests regarding a population median as discussed in your Text assume only that the population being sampled is _____ and that the sample is _____ .

A.2 A study has been undertaken to estimate the median research and development expenditures per firm for a population of industrial goods manufacturers. The value of the population median being estimated is denoted by _____

A.3 For the situation in A.2, a random sample of 75 manufacturers yields sample mean expenditures of $2.27 million and sample median expenditures of $0.46 million. These sample statistics are denoted _____ and _____ , respectively. Since the sample mean is so much larger than the sample median, it appears that this population is extremely skewed to the right. Therefore, the sample _____ is probably a more meaningful measure of the center of the distribution of expenditures than is the sample _____ .

A.4 A random sample of $\eta = 8$ households is selected from a population of households in a large suburban area to estimate the median income per household last year. In ascending order, the sample observations (in $ thousands) are 14.6, 14.9, 16.2, 16.9, 17.3, 18.7, 29.0 and 48.8. For this situation, Md = _____ .

A.5 For the situation in A.4, the following confidence interval was obtained: $14.9 \le \eta \le 29.0$. For this confidence interval, $r =$ _____ , $L_r =$ _____ , $U_r =$ _____ , and, from Table C-8 in your Text, $1 - \alpha =$ _____ .

A.6 For the situation in A.4, if a confidence coefficient near 99 percent had been desired, we would have used $r =$ _____ , $L_r =$ _____ , and $U_r =$ _____ , with $1 - \alpha =$ _____ . Hence, the appropriate confidence interval would be _____ $\le \eta \le$ _____ .

A.7 Suppose it is desired to estimate a population median from a random sample of $\eta = 49$ items with a 98 percent confidence interval. From (15.2) in your text, r is the largest integer which does not exceed $.5[n + 1 - z(1 - \alpha/2)\sqrt{n})]$ = _____ . Thus, r = _____, L_r is the _____th _____outcome in the sample and U_r is the _____th _____outcome in the sample.

A.8 A random sample of $\eta = 49$ accounts in a large accounts receivable file is selected to estimate η, the population median outstanding balance. Partial results are $L_1 = \$0$, ... $L_{15} = \$105$, $L16 = \$143$, $L_{17} = \$125$, ... $L_{32} = U_{18} = \$138$, $L_{33} = U_{17} = \$139$, $L_{34} = U_{16} = \$143$, $L_{49} = U_1 = \$683$. Using the results in A.7, the 98 percent confidence interval for η is _____ $\leq \eta \leq$ _____.

A.9 Explain why the alternative H_0: $\eta = 150$ is equivalent to the alternative H_0: $p = P(f_i > 150) = .5$ in the sign test for a population median. _____

A.10 In the sign test for a population median, explain what the symbol B denotes. _____

_____

A.11 Consider the alternatives H_0: $\eta \geq 20$ and H_1: $\eta < 20$ in a sign test for a population median. Alternative H_1 should be concluded only if B (or, equivalently, $p = B/N$ for a large sample test) is relatively _____. Explain why this is so.

_____

A.12 In a sign test involving the alternatives H_0: $\eta = 35$ and H_1: $\eta \neq 35$, a random sample of $\eta = 15$ observations is selected, of which nine are larger than 35.0. Hence the test statistic B = _____. In this situation, the exact sampling distribution of B when $\eta = n_0 = 35.0$ (i.e., when H_0 holds) is a(n) _____ probability distribution with $\eta =$ _____ and $p =$ _____.

A.13 If the sample size in a sign test is $\eta = 64$, the sampling distribution of the test statistic where $\bar{p} = B/N$ is _____ _____. $\sqrt{}$

$$z* = \frac{\bar{p} - p}{\sqrt{p_0(1-p_0)/n}}$$

»Problem 1

An analyst for a bank is studying account balances of daily
interest savings accounts with the bank. Since the analyst is aware
that the population of account balances is highly skewed, he has decided
that the population median balance per account, η, is a more meaningful
measure of the center of the distribution of account balances than is
the population mean balance per account, μ. A random sample of $\eta = 64$
of the bank's 9700 daily interest savings accounts is selected. Current
balances for there 64 accounts (in dollars) appear in ascending order
(row by row) in Table 15.1.

Table 15.1
ARRAY OF CURRENT ACCOUNT BALANCES (IN DOLLARS)
FOR A SAMPLE OF $\eta = 64$ DAILY INTEREST SAVINGS ACCOUNTS
(See Problem 1)

113	118	144	145	154	247	269	271
287	305	349	382	395	420	467	473
594	681	718	731	789	849	891	909
978	1014	1033	1043	1085	1145	1162	1190
1200	1252	1308	1363	1395	1412	1488	1535
1676	1800	1936	1989	2032	2043	2114	2163
2291	2572	2855	3085	3218	3656	3883	4247
4363	4702	5108	5341	6412	6872	7035	8326

a. i. The analyst wishes to use the sign test to determine whether
or not the population median balance per account is equal to
$2000. Conduct the appropriate test for η using $\alpha = .10$ when
$\eta=\eta_0=2000$. Specify the alternatives (in terms of both η and
p) and the decision rule, determine z^*, and state the
resulting decision. (Note: Since the observations in Table
15.1 are arrayed in ascending order, B can be determined by
simply counting the number of observations that exceed η_0. ✓

   ii. Explain in words how your test procedure would have
changed had the analyst been interested in testing
whether or not the population median balance per account
is greater than $2000 (i.e., H_0: $\eta \leq 2000$ and H_1: $\eta > 2000$).

b. Having concluded from the results in (a.i.) that $\eta \neq 2000$, the analyst wishes to estimate η with a 90 percent confidence interval. Construct and interpret the appropriate confidence interval. _____

_____ √

»Problem 2

A manager for a manufacturing company wished to estimate η, the median time (in hours) it takes new employees to complete a self-paced training unit in job safety. A random sample of fifteen new employees was Given the training unit. The employees' completion times are presented in Table 15.2.

Table 15.2
COMPLETION TIMES (IN HOURS) FOR THE SELF-PACED JOB SAFETY
TRAINING UNIT FOR SAMPLE OF $\eta = 14$ NEW EMPLOYEES

Employee	Time f_i	$f_i - \eta_0$	Sign	Employee	Time f_i	$f_i - \eta_0$	Sign
1	6.8	_____	_____	9	5.1	_____	_____
2	4.1	_____	_____	10	5.9	_____	_____
3	3.9	_____	_____	11	5.3	_____	_____
4	5.1	_____	_____	12	9.1	_____	_____
5	5.8	_____	_____	13	5.4	_____	_____
6	8.4	_____	_____	14	7.8	_____	_____
7	6.4	_____	_____	15	4.2	_____	_____
8	4.5	_____	_____				

a. i. Construct an array showing these 15 sample observations in ascending order.

ii. Obtain the sample median for these observations. Md = _____ √

b. Construct a confidence interval for η, the population median completion time for new employees, using a confidence coefficient near but not less than 95 percent. _____ √

c. i. Obtain the confidence interval if L_5 and U_5 had been selected as the confidence limits. _____

    ii. Would you expect the confidence coefficient associated with this interval to be smaller or larger than that associated with the interval in (b.)? _____ Explain.

_____ ✓

  d. According to a brochure furnished by a publisher of the training unit, one-half of the persons who use the unit complete it within 6.0 hours. The manager wishes to test whether of not η, the population median completion time for new employees, is equal to 6.0 hours, with α near but not greater than .05 when $\eta - \eta_0 - 6.0$. Conduct the appropriate test. Specify the alternatives (in terms of both η and p) and decision rule, determine B, and state the resulting decision. ✓

B.   INFERENCES ABOUT $n_1 - n_2$
INDEPENDENT SAMPLES, RANDOMIZED DESIGN
(Text: Chapter 15, Section 15.2)

Review of Basic Concepts

B.1 The magnitude by which two identically shaped population distributions differ in location is called the _____ _____ and is denoted by _____ .

B.2 If $\delta - 0$, the location of the population 1 distribution is _____ to the location of the population 2 distribution. In that case, it is also true that μ_1 _____ μ_2 and n_1 _____ n_2.

B.3 Explain in words how the sample statistic S_2, used in the M-W-W test, is computed. _____

_____

B.4 In a study undertaken to examine the relative locations of two distributions using the M-W-W test, $n_1 - 10$ and $n_2 - 15$. In this situation, the smallest possible value of S_2 is _____ and the largest possible value of S_2 is _____ .

B.5 For the situation in B.4, when $\delta - 0$ the expected value $E(S_2)$ of the sampling distribution of S_2 equals _____ and the standard deviation, $\sigma(S_2)$, equals _____ .

B.6 In applying the M-W-W test, the sampling distribution of S_2 when $\delta = 0$ is approximately normal provided $n_1 \geq$ _____ and $n_2 \geq$ _____.

B.7 If the alternative in the M-W-W test are of the form H_0: $\delta \geq 0$ and H_1: $\delta < 0$, alternative H_1 should be selected for relatively _____ values of S_2. √

»Problem 3

The regional manager for a chain of convenience grocery stores is attempting to determine on which of two major arterial roads a new store should be located. One component of this decision concerns traffic flow on these two roads. The city traffic engineer has provided the following recent traffic accounts for the two roads, listed in ascending order. Each count represents the total vehicular traffic for a 24-hour period.

$$\text{Division Street } X_1: \quad 1560 \quad 1571 \quad 1591 \quad 1623 \quad 1685$$
$$1750 \quad 1862 \quad 2069 \quad 2073 \quad 2122$$

$$\text{Tenth Street } X_2: \quad 1412 \quad 1486 \quad 1534 \quad 1618 \quad 1715$$
$$1732 \quad 1751 \quad 1830 \quad 1968 \quad 1973$$

Based on conversations with the traffic engineer, the regional manager is sure that the population distributions of traffic counts have the same shape and that the preceding traffic counts can be viewed as independent random samples of sizes $n_1 = 10$ and $n_2 = 10$.

a. Specify the alternatives for the M-W-W test of whether or not the two population distributions of traffic counts have the same location. _____ √

b. Specify the appropriate decision rule for this M-W-W test, with $\alpha = .10$, when $\delta = 0$. _____

c. Determine the value of S_2. _____ √

d. Determine the mean, variance, and standard deviation of the sampling distribution of S_2 when $\delta = 0$.

$E\{S_2\}$ = _____ ; $\sigma^2\{S_2\}$ = _____ ; $\sigma\{S_2\}$ = _____

e. What conclusion should be reached? _____
_____ √

C. INFERENCES ABOUT η_D
MATCHED SAMPLES, BLOCK DESIGN
(Text: Chapter 15, Section 15.3)

Review of Basic Concepts

C.1 When using matched samples to compare tow populations,
nonparametric procedures for single population studies can be
employed. This is so since the differences D = _____ in
matched samples constitute a random sample from a single population
of differences.

C.2 When comparing two populations based on matched samples, the
parameter which measures how far apart the two population
distributions are is denoted by _____, which is the
_____ of the population of differences.

C.3 Ten cities were selected for an experiment designed to measure the
impact of a new advertising campaign on sales of a particular
product. In this study, D was defined as sales volume in each city
one week after the new campaign had run (X_2) minus sales volume one
week before the campaign had run (X_1). In ascending order, the D
values are (in $ thousands) -.072, -.23, .34, .79, 1.26, 1,91,
2.25, 3.19, 3.21 and 3.40. To construct a confidence interval for
η_D with r = 3, the appropriate confidence coefficient from Table
C-8 is _____.   The resulting confidence interval is
_____ $\leq \eta_D \leq$ _____ .

C.4 Interpret the confidence interval in C.3. _____
_____

C.5 Tests concerning η_D can be conducted with either the sign test or
the Wilcoxon signed rank test. The sign test requires only that
the population of differences be _____ whereas the
Wilcoxon signed rank test requires that it be both _____
and _____. When both assumptions are satisfied, the
Wilcoxon signed rank test is generally more _____ than the
sign test. That is, when both tests have the same α risk, the
Wilcoxon signed rank test has a _____ probability of
concluding H_1 when H_1 is true.

C.6 For the situation in C.3, consider the alternatives H_0: $\eta_D \leq 0$ and H_1: $\eta_D > 0$. The value of the test statistic for this small sample test is B = _____. When $\eta_D = 0$, B follows a(n) _____ probability distribution with η = _____ and p = _____.

C.7 Consider a test for η_D involving the alternatives H_0: $\eta_D \geq 0$ and H_1: $\eta_D < 0$, with $\alpha = .01$ when $\eta_D = 0$. The sample data for this test were obtained from a random sample of $\eta = 200$ matched pairs. It cannot be assumed here that the population of differences D is symmetrical. Hence, this test must be conducted as a(n) _____ test. The appropriate decision rule in terms of the standardized test statistic z* is _____.

C.8 For the situation in C.7, 87 of the differences were found to be positive. Hence, the sample statistic p - _____, the test statistic z* = _____, and alternative _____ is concluded.

C.9 The test statistic T for the Wilcoxon signed rank test is defined as the _____ of the _____ associated with the D's.

C.10 For the situation in C.3, the first difference, -.72, has a signed rank value of _____. For the sample of $\eta = 10$ differences, we obtain T = _____.

C.11 For the situation in C.3, assume the population of differences is symmetrical. Then, the sampling distribution of T is _____ _____ with E(T) = _____, $\sigma^2(T)$ = _____, and $\sigma(T)$ = _____. √

»Problem 4

A family studies researcher has developed a questionnaire which can be scored on a continuum to provide an index of marital satisfaction of each partner in a marriage. The questionnaire is administered to a random sample of $\eta = 25$ married couples. For each couple, the difference D is defined as the score received by the wife (X_2) minus the score received by the husband (X_1). These sample differences appear in Table 15.3. It can be assumed that the population of differences is symmetrical.

Table 15.3
DIFFERENCES IN SCORES ON MARITAL SATISFACTION
QUESTIONNAIRE BETWEEN WIVES (X_2) AND HUSBANDS (X_1)
FOR SAMPLE OF η = 25 COUPLES
(See Problem 15.4)

Couple	Differences $D = X_2 - X_1$	Absolute Differences $|D|$	Rank	Signed Rank
1	16.8	16.8		
2	-17.0	17.0		
3	10.4	10.4		
4	8.9			
5	-7.4			
6	8.5			
7	1.1			
8	-8.3			
9	24.7			
10	17.9			
11	36.8			
12	-13.0			
13	3.3			
14	-0.4			
15	12.7			
16	10.1			
17	38.8			
18	22.0			
19	-11.7			
20	-9.4			
21	30.3			
22	-0.7			
23	19.2			
24	12.0			
25	-17.6			

a. State the appropriate alternatives for testing whether or not the population median difference in marital satisfaction score per couple is zero.

b. Construct the decision rule for conducting this test with the Wilcoxon signed rank test, using α = .10 when η_D = 0. _____
   _____√

c. Complete the computations in Table 15.3 and determine the values of T and z* for this test.

   T = _____ ;   z* = _____

d. What conclusions should be reached? _____
   Explain. _____
   _____ √

Answers to Chapter Reviews and Problems

A.1 continuous; random

A.2 η (eta)

A.3 $\overline{X}$; Md; median; mean

A.4 Md = 17.1

A.5 r-2; L_r - L_2 - 14.9; U_r - U_2 - 29.0; 1 - α - .930

A.6 r=1; L_r = L1 = 14.6; U_r = U_1 = 48.8; 1 - α = .992; 14.6 ≤ η ≤ 48.8

A.7 .5(49 + 1 - 2.326$\sqrt{49}$) = 16.859; r = 16; 16th smallest; 16th largest

A.8 109 ≤ η ≤ 143

A.9 If η - 150 and the sample is random, then the probability p that any sample observation f_i falls above 150 must be .5 according to the definition of the population median for a continuous population (i.e., one-half of the observations in the population are larger than 150). Thus, H_0: η - 150 and H_0: p - $P(f_i > 150)$ = .5 are equivalent.

A.10 B denotes the number of sample observations which are larger than the sample value η_0.

A.11 Small; if H_1: η < 20 is true, the probability p that any sample observation fi falls above η_0 = 20 must be less than .5. Thus, if H_1: η < 20 is true, then B, the number of sample observations which are larger than η_0 - 20 will be small. Since H_1 can be concluded only if the sample evidence clearly suggests that H_1 is true, alternative H_1: η < 20 here should be concluded only if B (or, equivalently, p - B/n) is relatively small.

A.12 B - 9; binomial; η - 15; p - .5

A.13 approximately normal

Problem 1

a. i. Alternatives: H_0: η = 2000;
H_1: $\eta \neq$ 2000, or
H_0: p = $P(f_i > 2000)$ = .5;
H_1: p = $P(f_i > 2000) \neq$ .5

Decision Rule: If $|z*| \leq$ 1.645, conclude $H_0(\eta$ - 2000);
If $|z*| >$ 1.645, conclude $H_0(\eta \neq$ 2000)

$z*$: B = 20; $\bar{p}$ = 20/64 = .3125

$\sigma(\bar{p})$ = $\sqrt{.5(1-.5)/64}$ = .0625 when p - p_0 = .5
(i.e., when η - η_0 - 2000)
Thus, $z*$ = (.3125 - .5)/.0625 = -3.000.

Decision: Since $|z*|$ = 3.000 > 1.645, conclude $H_1(\eta \neq$ 2000), the population median balance per account does not equal \$2000.

ii. In terms of p, the appropriate alternatives would have been
H_0: p = $P(f_i > 2000) \leq$ .5 and H_1:p = $P(f_i > 2000) >$ .5.
Thus, the sign test would have been constructed as a one-sided upper-tail test for p, with a decision rule based on z(1 - α) = 1.282.

b. .5(64 + 1 - 1.645$\sqrt{64}$) - 25.92; r - 25; L_{25} - 978; U_{25} - 1535; The 90 percent confidence interval for η is 978 $\leq \eta \leq$ 1535. Thus, we can state, with 90 percent confidence, that the population median balance per account is between \$978 and \$1535.

Problem 2

a. i. 3.9, 4.1, 4.2, 4.5, 5.1, 5.1, 5.3, 5.4, 5.8, 5.9, 6.4, 6.8, 7.8, 8.4, 9.1

ii. Md - 5.4

b. From Table C-8 in your Text, with η - 15 and r - 4 we obtain 1-α - .965. Thus, L_4 - 4.5, U_4 - 6.8 and the 96.5 percent confidence interval for η is 4.5 $\leq \eta \leq$ 6.8.

c. i. 5.1 $\leq \eta \leq$ 6.4

ii. Smaller; Since the confidence interval is narrower, we would
    expect the confidence coefficient to decrease. (Note: From
    Table C-10 in your Text, with η = 15 and r = 5 we obtain 1 -
    α = .882.)

d. Alternatives: H_0: η = 6.0;
                  H_1: $\eta \neq$ 6.0, or
                  H_0: p = $P(f_i > 6.0)$ = .5;
                  H_1: p = $P(f_i > 6.0) \neq$ .5

Decision Rule: Following the procedure illustrated in Figure
                 15.3 in your Text, when H_0 holds, the exact
                 sampling distribution of $\bar{p}$ is binomial with n=15
                 and p=.5. From Table C-8 in your text, we obtain
                 $P(f < 4)$ = .0176 and $P(f > 11)$ = .0176. Thus,
                 with α = .0352, the appropriate decision rule is:
                 If 4 ≤ B ≤ 11, conclude H_1(n = 6.0); If B < 4 or
                 B > 11, conclude $H_1(\eta_1 \neq 6.0)$

B:

i	f_i	f_i-6	Sign	i	f_i	f_i-6	Sign
1	6.8	0.8	+	9	5.1	-0.9	-
2	4.1	-1.9	-	10	5.9	-0.1	-
3	3.9	-2.1	-	11	5.3	-0.7	-
4	5.1	-0.9	-	12	9.1	3.1	+
5	5.8	-0.2	-	13	5.4	-0.6	-
6	8.4	2.4	+	14	7.8	1.8	+
7	6.4	0.4	+	15	4.2	-1.8	-
8	4.5	-1.5	-				

The number of plus signs is B = 5

Decision: Since 4 < B = 5 < 11, conclude $H_1(\eta$ = 6.0), the
             population median completion time for new
             employees is equal to 6.0 hours.

B.1 shift parameter; δ (delta)

B.2 identical; $\mu_1 = \mu_2$; $\eta_1 = \eta_2$

B.3 The $n_1 + n_2$ observations are placed in ascending order and assigned
ranks (smallest = 1 up through largest = $n_1 + n_2$). S_2 is then
computed as the sum of the ranks assigned to the n_2 sample
observations population 2.

B.4 $15(15+1)/2 = 120$; $10(15) + 15(15+1)/2 = 270$

B.5 $E\{S_2\} = 15(10+15+1)/2 = 195$; $\sigma\{S_2\} = \sqrt{10(15)(10+15+1)/12} = 18.03$

B.6 10; 10

B.7 small

Problem 3

a. H_0: $\delta = 0$; H_1: $\delta \neq 0$

b. If $\left|z*\right| \le 1.645$, conclude $H_0(\delta = 0)$;
   If $\left|z*\right| > 1.645$, conclude $H_0(\delta \neq 0)$;

c. The combined array appears as follows (X_2 ranks boxed).

Value:	1412	1486	1534	1560	1571	1591	1618
Rank:	1	2	3	4	5	6	7
Value:	1623	1685	1715	1732	1750	1751	1830
Rank:	8	9	10	11	12	13	14
Value:	1862	1968	1973	2069	2073	1830	
Rank:	15	16	17	18	19	20	

Thus, $S_2 = 1 + 2 + 3 + 7 + 10 + 11 + 13 + 14 + 16 + 17 = 94$

d. $E\{S_2\} = 10(10+10+1)/2 = 105$
   $\sigma^2\{S_2\} = 10(10)(10+10+1)/12 = 175$

   $\sigma\{S_2\} = \sqrt{175} = 13.23$

e. $z* = (94 - 105)/13.23 = -.831$; Since $\left|z*\right| = .831 < 1.645$, conclude $H_0(\delta = 0)$, the two populations of traffic counts have the same location.

C.1 $D = X_2 - X_1$

C.2 η_D; median

C.3 .891; $.34 \le \eta_D \le 3.19$

C.4 with 89.1 percent confidence, we can state that the population
median difference in sales volume per city one week after the
campaign compared to one week before the campaign is between $.34
thousand and $3.19 thousand.

C.5 continuous; continuous and symmetrical; powerful; higher

C.6 B = 8 (i.e., the number of positive sample differences); binomial;

$\eta = 10$; $p = .5$

C.7 sign; If $z* \leq -2.326$, conclude $H_0(\eta_D \geq 0)$; If $z* < -2.326$, conclude
$H_1(\eta_D < 0)$

C.8 $\bar{p} - 87/200 - .435$; $z* - (.435-.500)/\sqrt{.5(1-.5)/200} - -1.838$; H_0

C.9 sum of the signed ranks

C.10 -3; $T = -3 - 1 + 2 + 4 + 5 + 6 + 7 + 8 + 9 + 10 = 47$

C.11 approximately normal; $E(T) = 0$; $\sigma^2(T) = 10(11)(21)/6 = 385$;
$\sigma(T) = 19.62$

Problem 4

a. H_0: $\eta_D - 0$; H_1: $\eta_D \neq 0$

b. If $|z*| \leq 1.645$, conclude $H_0(\eta_D = 0)$;
   If $|z*| > 1.645$, conclude $H_0(\eta_D \neq 0)$

c.

i	$|D|$	Rank	Signed Rank	i	$|D|$	Rank	Signed Rank
1	16.8	16	16				
2	17.0	17	-17	14	0.4	1	-
3	10.4	11	11	15	12.7	14	14
4	8.9	8	8	16	10.1	10	10
5	7.4	5	-5	17	38.8	25	25
6	8.5	7	7	18	22.0	21	21
7	1.1	3	3	19	11.7	12	-12
8	8.3	6	-6	20	9.4	9	9
9	24.7	22	22	21	30.3	23	23
10	17.9	19	19	22	0.7	2	2
11	36.8	24	24	23	19.2	20	-20
12	13.0	15	-15	24	12.0	13	13
13	3.3	4	4	25	17.6	18	-18

$$T = 155$$

$E(T) = 0$

$\sigma(T) = \sqrt{25(26)(51)/6} = 74.33;$
Thus, $z* = (155 - 0)/74.33 = 2.085$

d. Since $|z*| = 2.095 > 1.645$, conclude $H_1(\eta_D \neq 0)$, the population median difference in marital satisfaction score per couple does not equal zero.

16

GOODNESS OF FIT

A. GOODNESS OF FIT TESTS AND CHI-SQUARE TESTS INVOLVING DISCRETE AND CONTINUOUS PROBABILITY DISTRIBUTIONS

(Text: Chapter 16, Sections 16.2 to 16.4)

Review of Basic Concepts

A.1 The alternative H_0 in a goodness of fit test specifies the
_____ _____ of the distribution. In addition, H_0
may also specify the values for the _____ of the
distribution.

A.2 In the chi-square goodness of fit test, the symbol f_i is used to
denote the _____ frequency in the ith class. The
symbol F_i is used to denote the _____ frequency in
the ith class if _____ holds.

A.3 In a statistical study designed to test whether or not a random
sample came from a Poisson distribution with $\lambda = .40$, the
appropriate alternatives are H_0: _____ and H_1:
_____ .

A.4 For the situation in A.3, n = 60 observations resulted in the
following:

Class	Number of Observations
0	38
1	18
2 or more	4

In this situation, f_2 = _____ From Table C.6 in your
Text, the probability that an observation will fall into the second
class if the population is Poisson with $\lambda = .4$ is _____ .
Thus, F_2 = _____ . Further, the difference $f_2 - F_2$ =
_____ is called a(n) _____ .

A.5 For a chi-square goodness of fit test with k classes, when n is reasonably large and m parameters are estimated, the test statistic X^2 is distributed approximately as a(n) _____ random variable with _____ degrees of freedom when H_0 holds.

A.6 In order to use the chi-square approximation in A.5, the expected frequency in each class should be _____ or more, and at least 50 percent of the classes should have an expected frequency of _____ or more.

A.7 For the situation in A.3 and A.4, the expected frequencies are 40.218, 16.086, 3.696. In this situation, is n large enough for the chi-square approximation to hold? _____ Explain.

_____

A.8 For the situation in A.3 and A.4, if f_2 - 18 and F_2 = 16.086 , then the appropriate relative squared residual for the second class, which is denoted by _____ ,equals _____ . If the relative squared residuals for the first and third classes are .1223 and .0250, respectively, the X^2 _____ .

A.9 For any pair of alternatives in a chi-square goodness of fit test, relatively large values of X^2 are indicative that alternative _____ is true.

A.10 Determine each of the following chi-square distribution percentiles from Table C-2 in your Text.

    a. $X^2(.90; 9)$ = _____
    b. $X^2(.95; 4)$ = _____
    c. $X^2(.99; 1)$ = _____
    d. $X^2(.995; 7)$ = _____

A.11 Determine the value of the chi-square distribution percentile that forms the boundary between the acceptance and rejection regions for each of the following tests. For each, X^2 is computed on the basis of observations classified into 6 classes.

    a. Test of whether or not a distribution is normal, given α = .025.

_____

   b. Test of whether or not a distribution is normal with $\mu = 0$ and
      $\sigma = 1$, given $\alpha = .025$. _____

   c. Test of whether or not a distribution is Poisson, given $\alpha = .01$.
      _____

   d. Test of whether or not a distribution is Poisson with $\lambda = 12$,
      given $\alpha = .01$. _____

A.12 It is desired to test whether or not a certain probability
distribution is Poisson with $\lambda = 0.5$. A random sample of $n = 100$
items is to be used in conducting the test. The largest number of
classes into which the sample observations can be classified
without violating the working rule given by (16.7) in your Text is
_____. Identify these classes and their respective expected
frequencies. _____
_____

A.13 For continuous distributions, the discriminating capability of the
chi-square test is increased if, for a given number of classes, the
class intervals are chosen such that the classes have
_____ expected frequencies.

A.14 A chi-square test is being conducted to determine whether or not a
distribution is normal. Sample results are $n = 50$, $X = 10.0$, and
$s = 2.0$. Ten classes are to be used. The second lowest class will
consist of X values of magnitude at least _____ but less
than _____.

A.15 For the situation in A.14, $F_2 =$ _____. $\sqrt{}$

A.16 Consider the following alternatives in a chi-square test with
$\alpha = .05$ when H_0 holds.

   H_0: The probability distribution is normal with $\mu = 15.0$ and
      $\sigma = 2.5$

   H_1: The probability distribution is not normal with $\mu = 15.0$
      and $\sigma = 2.5$

If alternative H_1 is selected here, is it then concluded that the probability distribution is not normal? _____ Explain.

_____

_____

A.17 For the situation in A.16, a computer package used to aid in conducting this test gave the upper-tail P-value for the test as .0072. Thus, alternative _____ is concluded. √

»**Problem 1**

An office manager was involved in a study of the use of five statewide, toll-free telephone lines assigned to his office. One aspect of this study was to determine the distribution of X, the number of phone lines in use at any time during working hours. At his request, the company operator noted the number of lines in use at n = 40 randomly selected times during working hours. These observations appear in Table 16.1a.

Table 16.1
CHI-SQUARE GOODNESS OF FIT TEST
FOR STUDY OF TELEPHONE LINE USE
(See Problem 1)

(a)

Number of Lines in Use at 40 Randomly Selected Times

1	3	1	0	0	2	0	1	3	2
3	1	1	2	2	4	0	2	1	1
0	1	5	0	1	2	3	1	0	2
0	0	3	2	0	1	3	1	4	1

(b)
Calculation of X^2 for Test of Whether or Not
the Distribution is Discrete Uniform with a = 0 and s = 6

Class i	Number of PhoneLines in Use X	Observed Frequency f_i	Probability Under H_0	Expected Frequency Under H_0 F_i	$f_i - F_i$	$\dfrac{(f_i - F_i)^2}{F_i}$
1	0	.10	.1667	6.663	3.332	1.665
2	1	___	___	___	___	___
3	2	___	___	___	___	___
4	3	___	___	___	___	___
5	4	___	___	___	___	___
6	5	___	___	___	___	___
	Total	___	___	___	___	___

a. The manager's first thought was that, at any point in time, it was equally likely that 0, 1, 2, 3, 4, or all 5 lines would be in use. State this belief in the form of alternatives concerning the nature of the probability distribution of X.

_____ √

b. Specify the appropriate decision rule for this chi-square goodness of fit test, using α = .01 when H_0 holds.

_____
_____ √

c. Complete the computations in Table 16.1b to obtain the test statistic X^2 for this test. X^2 = _____

d. What conclusion should be reached about the distribution of number of lines in use at any point in time? _____
_____ √

e. After concluding the X does not follow the discrete uniform distribution with a = 0 and s = 6, the manager consulted a research analyst who suggested that X should follow a Poisson distribution. Conduct this chi-square goodness of fit test, given α = .01 when H_0 holds. Specify the alternatives and decision rule, determine X^2 and state the resulting decision. (Note: The mean of the 40 observations is X = 1.50. √

»Problem 2

The lifetimes, in hours, of a random sample of n = 24 forty-watt incandescent lamps in quality control forced-life tests appears in Table 16.2a. For these 24 observations, $\bar{X}$ = 1,153.0 and s = 175.2.

Table 16.2
CHI-SQUARE GOODNESS OF FIT TEST
FOR STUDY OF LIFETIMES OF INCANDESCENT LAMPS
(See Problem 2)

(a)
Lifetimes of Sample of 24 Lamps

1,105	1,243	1,204	1,203	1,310	1,262	1,234	1,104
1,303	1,185	769	1,404	944	1,343	932	1,055
1,381	816	1,067	1,252	1,248	1,324	1,000	984

(b)
Calculation of X^2 for Test of Whether or Not
the Distribution of Lifetimes is Normal

Class i	Class Intervals For Lifetimes X	Observed Frequency f_i	Probability Under H_0	Expected Frequency Under H_0 F_i	$f_i - F_i$	$(f_i - F_i)^2 / F_i$
1	___	___	___	___		___
2	___	___	___	___	___	___
3	___	___	___	___	___	___
4	___	___	___	___	___	___
	Total	___	___	___	___	___

a. State the appropriate alternatives for testing whether or not the life-times of the lamps are normally distributed.

b. Determine the class intervals for classifying the 24 sample observations, given that four classes are to be used. Insert these class intervals in Table 16.2b. √

c. Specify the decision rule for this test, using α = .05 when H_0 holds.

d. Complete the computations in Table 16.2b to obtain the value of X^2 for this test. X^2 = _____

e. On the basis of this test, what conclusion should be reached about the distribution of lifetimes of the lamps? _____

_____ √

Answers to Chapter Reviews and Problems

A.1 functional form; parameters

A.2 observed; expected; H_0

A.3 H_0: The probability distribution is Poisson with λ = .4; H_1: The probability distribution is not Poisson with λ = .4

A.4 f_2 = 18; P(2) = .2681; F_2 = 60(.2681) = (16.086); f_2 - F_2 = 1.914; residual

A.5 chi-square (X^2); k - m - 1

A.6 two; five

A.7 Yes; the expected frequency in each class is 2 or more, and the expected frequency in two of the three classes (i.e, in 67 percent of them) is 5 or more.

A.8 $(f_2-F_2)^2/F_2$ = (18-16.086)/16.086 = .2277; X^2 = .3750

A.9 H_1

A.10 a. X^2(.90; 9) = 14.68

b. X^2(.95; 4) = 9.49

c. X^2(.99; 11) = 24.73

d. X^2(.995; 7) = 20.28

A.11 a. A = X^2(.975; 6-2-1) = 9.35

b. A = X^2(.975; 6-0-1) = 12.83

c. $A = X^2(.99; 6-1-1) = 13.28$

d. $A = X^2(.99; 6-0-1) = 15.09$

A.12 3; X:

	0	1	2 or more
F_i:	60.65	30.33	9.02

A.13 equal

A.14 $\overline{X} + z(.10)s = 10 - 1.282(2) = 7.44$; $\overline{X} + z(.20)s = 10 - .84(2) = 8.32$

A.15 $.10(50) = 5.0$

A.16 No; It is concluded that either the distribution is not normal or it is normal but at least one of the parameters (i.e., μ and/or σ) has a value other than that specified in H_0.

A.17 H_1

Problem 1

a. H_0: The probability distribution is discrete uniform with $a = 0$ and $s = 6$;

H_1: The probability distribution is not discrete uniform with $a = 0$ and $s = 6$.

(Note: a (smallest possible outcome) and s (number of distinct possible outcomes are the parameters of the discrete uniform distribution. For a review of this distribution, refer to Chapter 6, Section 6.3 in your Text.)

b. $k = 6$; $m = 0$; $1 - \alpha = .99$; $X^2(.99; 5) = 15.09$. Thus, the decision rule is:

If $X^2 \leq 15.09$, conclude H_0;
If $X^2 > 15.09$, conclude H_1.

c.

i	X	f_i	Probability Under H_0	F_i	$f_i - F_i$	$(f_i - F_i)^2 / F_i$
1	0	10	.1667	6.668	3.332	1.665
2	1	13	.1667	6.668	6.332	6.013
3	2	8	.1667	6.668	1.332	.266
4	3	6	.1667	6.668	-.668	.067
5	4	2	.1667	6.668	-4.668	3.268
6	5	1	.1667	6.668	-5.668	4.818
	Total	40	1.00	40.0	0.0	X^2=16.097

d. Since X^2 = 16.097 > 15.09, conclude H_1, the probability distribution is not discrete uniform with a = 0 and s = 6.

e. Alternatives: H_0: The probability distribution is Poisson; H_1: The probability distribution is not Poisson

Decision Rule: k = 5 (Note: As shown in the computation of X^2 to follow, the class "4 or more" must be formed by pooling so that each F_i is 2 or more. Hence, k = 5); m = 1; X^2(.99; 3) = 11.34. Thus, the decision rule is:

If $X^2 \le$ 11.34, conclude H_0;
If $X^2 >$ 11.34, conclude H_1

X^2:

i	X	f_i	Probability Under H_0	F_i	$f_i - F_i$	$(f_i - F_i)^2 / F_i$
1	0	10	.2231	8.924	1.076	.1297
2	1	13	.3347	13.388	-.388	.0112
3	2	8	.2510	10.040	-2.040	.4145
4	3	6	.1255	5.020	.980	.1913
5	≥4	3	.0657	2.628	.372	.0527
	Total	40	1.00	40.0	0.0	X^2=.7994

Decision: Since X^2 = .7994 < 11.34, conclude H_0, the Poisson probability distribution provides an adequate fit.

Problem 2

a. H_0: The probability distribution is normal;
   H_1: The probability distribution is not normal.

b. We need the 25th, 50th, and 75th percentiles of the normal distribution using $\overline{X}$ = 1153 and s = 175.2. These are obtained as follows:

   25th percentile: 1153 - .67(175.2) = 1035.6
   50th percentile: 1153 +   0(175.2) = 1153.0
   75th percentile: 1153 + .67(175.2) = 1270.4

   The resulting class intervals are: Under 1035.6; 1035.6 - 1152.9; 1153.0 - 1270.3; 1270.4 and over.

c. k = 4; m = 2; 1 - α = .95; X^2(.95; 1) = 3.84. Thus, the decision rule is:

   If $X^2 \leq 3.84$, conclude H_0;
   If $X^2 > 3.84$, conclude H_1.

d.

i	X	f_i	Probability Under H_0	F_i	f_i-F_i	$(f_i$-$F_i)^2/F_i$
1	< 1035.6	6	.25	6	0	0
2	1035.6-1152.9	4	.25	6	-2	.67
3	1153.0-1270.3	8	.25	6	2	.67
4	≥ 1270.4	6	.25	6	0	0
	Total	24	1.00	24	0	X^2=1.34

e. Since X^2 = 1.34 < 3.84, conclude H_0, the normal probability distribution provides an adequate fit.

17

MULTINOMIAL POPULATIONS

A. INFERENCES CONCERNING THE PARAMETERS p_i OF MULTINOMIAL POPULATIONS

(Text: Chapter 17, Sections 17.1 to 17.3)

Review of Basic Concepts

A.1 In a statistical study of the order-filling process in a warehouse, the status of each order is classified as (1) waiting to be processed, in-process, (3) on back-order, or (4) in-transit. Thus, the number of different attribute categories for each order is _____ which is denoted by the symbol _____.

A.2 For the situation in A.1, the population of orders is multinomial since each order is classified into only _____ of the attribute categories and there are at least _____ such categories.

A.3 For the situation in A.1, explain what the parameter p_3 specifies.

_____

A.4 A construction division of a large government agency is responsible for numerous construction projects. The probabilities for projects being in three major project stages are as follows.

i	Project Stage	Probability p_i
1	Negotiation	0.5
2	Planning	0.3
3	Construction	0.2
	Total	1.0

A random sample of n = 5 projects is to be selected. The multinomial probability function for this sampling situation involves four parameters. They are _____, _____, _____ and _____.

A.5 For the situation in A.4, it is desired to determine the probability that the n = 5 projects will be distributed as follows: two in negotiation, one in planning and two in construction. Thus, it is desired to determine the probability that f_1 = _____ , f_2 = _____ , and f_3 = _____ . This probability, which is denoted _____ , can be calculated from the expression _____ , which equals _____ .

A.6 The multinomial probability function (17.2) in your Text yields exact probabilities when the size of the population is _____ . It yields approximations of the probabilities when the size of the population is _____ , so long as the sampling fraction n/N is _____ .

A.7 For the situation in A.1, it is desired to construct a 95 percent confidence interval for p_4, the probability that an order is in-transit, based on a random sample of orders. Explain why this can be constructed as a confidence interval for a population proportion as described in Chapter 13 of your Text. _____

_____

A.8 The population of interest in a study consists of all households in a northern state. Each household is classified according to whether its primary heating fuel is oil, natural gas, or other. It is desired to test whether the parameters of this multinomial population are .70, .25, and .05, respectively. The alternatives for this test are H_0: _____

and H_1: _____ .

A.9 For the situation in A.8, does alternative H_1 necessarily imply that $p_1 \neq .70$? _____ Explain. _____

_____

A.10 For the situation in A.8, a random sample of n = 1,000 households was selected, and a computer package was used to aid in conducting the test. The computer package gave the P-value for the test as .2456. If the risk specification is α = .10 when H_0 holds, alternative _____ is concluded. √

»Problem 1

Based on a recent census, it is known that the distribution of housing units by type of structure in a large midwestern state is as follows: single unit structures = 78%; structures of two to nine units = 7%; structures of ten or more units = 15%. A cabinet wholesaler serving a part of the state has recently surveyed a random sample of n = 800 of the estimated 90,000 housing units in his market. The following results were obtained: single unit structures = 696; structures of two to nine units = 28; structures of ten or more units = 76.

a. State the appropriate alternatives for testing whether the distribution of housing units by type of structure in the wholesaler's market is the same as that for the entire state.

_____

b. Specify the decision rule for this test, using α = .10 when H_0 holds. _____ $\checkmark$

c. Compute the X^2 statistic by completing the following table.

Units In Structure	Class i	Observed Frequency f_i	Probability Under H_0 p_{i0}	Expected Frequency Under H_0 $F_i = np_{i0}$	$f_i - F_i$	$(f_i - F_i)^2 / F_i$
1	1	696	.78	624	72	8.308
2-9	2	28	_____	_____	_____	_____
10 or more	3	76	_____	_____	_____	_____
Total		800	_____	_____	_____	$X^2 =$ _____

d. Is the sample size in this study large enough so that X^2 is distributed approximately as a X^2 random variable with k - 1 degrees of freedom under H_0? _____ Explain. _____

_____

e. What conclusion should be reached? _____ Explain.

_____

_____ $\checkmark$

f. Based on the residuals f_i-F_i, how does the distribution of housing units in the wholesaler's market appear to differ from that in the state as a whole? _____

_____

g. Construct a 90 percent confidence interval for p_1, the proportion of housing units in single unit structures in the wholesaler's market. _____ ✓

B. BIVARIATE MULTINOMIAL POPULATIONS
(Text: Chapter 17, Section 17.4)

Review of Basic Concepts

B.1 Identify whether or not each of the following is a bivariate multinomial population and explain why or why not.

a. Residents of a metropolitan area are classified as either owning or renting their place of residence. _____

_____

b. Employees of a company are simultaneously classified by gender and by marital status. _____

_____

B.2 The population of private homes in a metropolitan area is distributed by age and type of home according to the following bivariate multinomial probability distribution.

		Age of Home	
		j=1	j=2
Type of Home		Ten Years or Less	More than Ten Years
i=1	Ranch	.12	.42
i=2	Split Level	.15	.07
i=3	Other	.03	.21

Give the value and interpret the meaning of each of the following for this population.

a. P_{12} = _____

b. P_{31} = _____

    c. $p_{2.}$ = _____

    d. $p_{.1}$ = _____

B.3 For the situation in B.2, p_{12} and p_{31} are called _____ probabilities while $p_{2.}$ and $p_{.1}$ are called _____ probabilities.

B.4 Are the two variables in B.2 statistically independent? _____ Explain. _____

_____

B.5 For the situation in B.2, obtain the conditional probability distribution for type of home among those homes which are ten years old or less.

Type of Home	Probability Conditioned Upon Ten Years Old or Less
Ranch	_____
Split Level	_____
Other	_____
Total	1.00

B.6 A random sample of n = 800 manufactured parts is simultaneously classified by quality level and machine used for manufacture. The following results are obtained.

	Machine		
	j=1	j=2	j=3
Quality Level	A	B	C
i=1 Defective	32	15	13
i=2 Not Defective	368	235	137

Identify each of the following sample frequencies.

    a. f_{11} = _____ ; f_{13} = _____ ; f_{22} = _____

    b. $f_{.1}$ = _____ ; $f_{.2}$ = _____ ; $f_{.3}$ = _____

    c. $f_{1.}$ = _____ ; $f_{2.}$ = _____ ; n = _____

B.7 For the situation in B.6, determine each of the following estimated probabilities.

a. $\bar{p}_{1.}$ = _____ ; $\bar{p}_{2.}$ = _____

a. $\bar{p}_{.1}$ = _____ ; $\bar{p}_{.2}$ = _____ ; $\bar{p}_{.3}$ = _____

B.8 For the situation in B.6, determine each of the following estimated probabilities p_{ij} and expected frequencies F_{ij}, under the assumption that machine and quality level are statistically independent.

a. $\bar{p}_{11}$ = _____ ; $\bar{p}_{13}$ = _____ ; $\bar{p}_{22}$ = _____

b. F_{11} = _____ ; F_{13} = _____ ; F_{22} = _____

B.9 A random sample of n = 300 households which bake cakes is classified by income class and whether they usually purchase cake mix or mix their own ingredients. The following table shows both the sample frequencies and the expected frequencies under the assumption that the two variables are statistically independent.

		Income ($ thousands)			
		j=1 Under 10	j=2 10-Under 20	j=3 20-Under 30	j=4 30 or More
i = 1	Purchase	24 (32.67)	100 (98.00)	85 (81.67)	36 (32.67)
i = 2	Purchase	16 (7.33)	20 (22.00)	15 (18.33)	4 (7.33)

Verify that F_{11} = 32.67. _____

_____

B.10 For the situation in B.9, explain in words how the X^2 statistic for testing the statistical independence of two variables is computed.

_____

_____

B.11 For the situation in B.9 and B.10, r = _____ and c = _____. Thus, with α = .01 when H_0 holds, the value of the chi-square distribution percentile that forms the boundary between the acceptance and rejection regions in the decision rule is

_____ .

B.12 A random sample of n = 600 employees with health insurance is selected from a very large population of such employees. Each is classified by type of health insurance plan - prepaid (i.e., a health maintenance organization) versus not prepaid - and by how often the employee receives a routine physical exam. The following results are obtained.

| | Type of Health Insurance Plan | |
| | j=1 | j=2 |
Quality Level	Prepaid	Not Prepaid
i=1 Every Year or More	143	153
i=2 Every Two to Five	78	139
i=3 Not Defective	39	48

It is desired to test whether or not the two variables are statistically independent, using α = .10 when H_0 holds. A computer package gave the P-value for this test as .0606. What conclusion should be reached? _____

B.13 For the situation in B.12, convert the observed frequencies to estimated probabilities of frequency of physical exam conditioned on type of insurance. What is the nature of the statistical dependence between the two variables? _____

_____

B.14 The bivariate distribution shown in B.12 is also referred to as a(n) _____ table. Thus, the test for statistical - independence is sometimes called a(n) _____ _____ test. √

»Problem 2

A random sample of 300 managers in middle-level managerial positions is classified by age class and education level. The results are shown in the following table, with education level measured as the highest college degree earned. It is desired to test whether or not age class and education level are statistically independent.

		Age Class		
Education Level	Under 35	35-Under 50	50 and Above	Total
No College Degree	4 (____)	30 (____)	26 (____)	60
Baccalaureate	35 (____)	115 (____)	60 (____)	210
Advanced Degree	21 (____)	5 (____)	4 (____)	30
Total	60	160	90	300

a. State the appropriate alternatives for this test. _____
   _____

b. Specify the decision rule for this test, using $\alpha = .01$ when H_0
   holds. _____ √

c. Calculate the expected frequencies under H_0 and insert them in
   the preceding table. √

d. Calculate the appropriate test statistic. _____

e. What conclusion should be reached? _____ Explain. _____
   _____

f. i. Construct the estimated conditional probability distribution
      for education level conditioned upon age class.

  ii. What seems to be the nature of the relationship between these
      variables? _____
      _____
      _____ √

C. COMPARISONS OF SEVERAL MULTINOMIAL POPULATIONS

(Text: Chapter 17, Section 17.5)

Review of Basic Concepts

C.1 Identify whether each of the following involves two or more
univariate multinomial populations or a bivariate multinomial
population.

    a. A random sample of students at a university is selected and classified on the basis of gender and academic major.

    _____

    b. A random sample of male students at a university and an independent random sample of female students at the same university are selected and each classified on the basis of academic major. _____

    c. Independent random samples of households in three metropolitan areas are selected and each household is classified as a nonuser or a user of a particular product. _____

C.2 For the situation in C.1(c.), there are c = _____ multinomial populations and r = _____ attribute classes in each population.

C.3 The probability that a randomly selected element from the second multinomial population is classified into the first attribute class is denoted by _____ .

C.4 Explain what each of the following denotes in the context of sampling from several multinomial populations.

    a. f_{12} _____

    b. n_2 _____

    c. n_T _____

    d. $\bar{p}_1 = f_{1.}/n_T$ _____

    e. $F_{12} = n_2 \bar{p}_1$ _____

C.5 Independent random samples of 300 male and 400 female teenagers are selected and classified on the basis of brand of cola most often purchased. Sample results appear in the following table.

| | Gender | | |
| | (j=1)
Male | (j=2)
Female | Total |
Brand			
(i=1)A	185	80	265
(i=2)B	95	190	285
(i=3)C	20	130	150
(i=4)D	300	400	700

Identify each of the following for this study.

a. f_{12} = _____ ; f_{21} = _____ ; f_{31} = _____

b. n_1 = _____ ; n_2 = _____ ; n_T = _____

c. $\bar{p}_1$ = _____ ; $\bar{p}_2$ = _____ ; $\bar{p}_3$ = _____

d. F_{12} = _____ ; F_{21} = _____ ; F_{31} = _____ ✓

»Problem 3

Refer to the situation in Problem 1. The cabinet wholesaler is considering expansion to a second market in a nearby state. As part of his assessment of this new market, an independent random sample of n = 500 of the 120,000 housing units in this market area is selected. The following results are obtained: single unit structures = 432; structures of two to nine units = 19; structures of ten or more units = 49.

a. State the appropriate alternatives to test whether or not the distribution of housing units by type of structure in the proposed new market is identical to that in the current market.

_____
                                                                    ✓

b. Specify the decision rule for this test, using α = .10 when H_0 holds. _____
                                                                    ✓

c. The sample frequencies for this test appear in the following table. Calculate the expected frequencies F_{ij} when H_0 holds and insert them in the table.

| | Market | | |
Units in Structure	Current	Proposed	Total
1	696 (_____)	432 (_____)	1,128
2-9	28 (_____)	19 (_____)	47
10 or more	76 (_____)	49 (_____)	125

$$n_1 = 800 \qquad n_2 = 500 \qquad n_T = 1,300$$

d. Calculate X^2. _____ ✓

e. What conclusion should be reached about the two multinomial
populations in this study? _____

_____

f. Would you anticipate that the estimated probabilities of type of
structure for the current market and the proposed market will
differ sharply from one another? _____ Explain. _____
_____ ✓

Answers to Chapter Reviews and Problems

A.1 4; k

A.2 one; at least two (in this case, there are four)

A.3 p_3 specifies the probability that an order selected at random is on
back-order.

A.4 $p_1 = 5$; $p_2 = .3$; $p_3 = .2$; $n = 5$

A.5 $f_1 = 2$; $f_2 = 1$; $f_3 = 2$; $P(2, 1, 2) = [5!/2!1!2!](.5)^2(.3)^1(.2)^2 = .090$

A.6 infinite; finite; small

A.7 To estimate p_4, the remaining three attribute categories can be
condensed to a single category, not in-transit, with probability $p_1
+ p_2 + p_3 = 1 - p_4$. Thus, for purposes of this inference, we have
a binomial situation and, hence, can use the inferential procedure
from Chapter 13 to construct a 95 percent confidence interval for
p_4.

A.8 H_0: $p_1 = p_10 = .70$; $p_2 = p_{20} = .25$ and $p_3 = p_{30} = .05$; H_1: $p_1 \neq p_{io}$ for some $i (i = 1, 2, 3)$.

A.9 No; H_1 implies only that not all p_i are as specified under H_0. The parameters under H_1, e.g., would be $p_1 = .70$, $p_2 = .10$, and $p_3 = .20$.

A.10 H_0

Problem 1

a. H_0: $p_1 = p_{10} = .78$, $p_2 = p_{20} = .07$, and $p_3 = p_{30} = .15$
H_1: $p_i \neq p_{io}$ for some $i (i=1, 2, 3)$

b. $1 - \alpha = .90$; $k - 1 = 2$; $X^2(.90;2) = 4.61$.
Thus, the decision rule is:
If $X^2 \leq 4.61$, conclude H_0; If $X^2 > 4.61$, conclude H_1.

c.

i	f_i	P_{io}	$F_i = np_{io}$	$f_i - F_i$	$(f_i - F_i)^2/F_i$
1	696	.78	624	72	8.308
2	28	.07	56	-28	14.000
3	76	.15	120	-44	16.133
Total 800		1.00	800	0	$X^2 = 38.441$

d. Yes. Since all F_i are at least 2 and 50 percent or more (in this case, all) are at least 5, the X^2 approximation holds.

e. H_1. Since $X^2 = 38.441 > 4.61$, conclude H_1, that the distribution of housing units by type of structure in this market does not follow the distribution for the state.

f. Since $f_1 - F_1 = 72$, it appears that a larger proportion of housing units in the wholesaler's market are single unit structure as compared to the state. Conversely, the proportions of housing units in both multi-unit structure classes appear to be smaller in the wholesaler's market than in the state, based on the residuals $f_2 - F_2 = -28$ and $f_3 - F_3 = -44$.

g. $\bar{p} = 696/800 = .87$; $s(\bar{p}_1) = \sqrt{.87(1-.87)/(800-1)} = .0119$; $z(.95) = 1.645$. The confidence limits are $.87 \pm 1.645(.0119)$. Thus, the 90 percent confidence interval for p_1 is $.850 \leq p_1 \leq .890$.

B.1 a. No; this is a univariate multinomial population with $k = 2$ attribute classes.

b. Yes; this multinomial population is bivariate since the attribute classes (i.e., combinations of gender and marital status) correspond to a bivariate classification system.

B.2 a. $p_{12} = .42$ is the proportion of homes which are both ranch style and more than ten years old.

b. $p_{31} = .03$ is the proportion of homes which are both of a style other than ranch or split level and ten years old or less.

c. $p_{2.} = .22$ is the proportion of homes which are split level style.

d. $p_{.1} = .30$ is the proportion of homes which are ten years old or less.

B.3 joint; marginal

B.4 No. If they were independent, $p_{ij} = p_{i.}p_{.j}$ for all i and j. Since this does not hold for all i and j (e.g., $p_{11} = .12 \neq .162 = p_{1.}p_{.1}$), the two variables are not statistically independent.

B.5

Type of Home	Probability Conditioned Upon Ten Years Old or Less
Ranch	.12/30 = .40
Split Level	.15/30 = .50
Other	.03/30 = .10
Total	1.00

B.6 a. $f_{11} = 32$; $f_{13} = 13$; $f_{22} = 235$

b. $f_{.1} = 400$; $f_{.2} = 250$; $f_{.3} = 150$

c. $f_{1.} = 60$; $f_{2.} = 740$; $n = 800$

B.7 a. $\overline{p_{1.}} = .075$; $\overline{p_{2.}} = .925$

b. $\bar{p}_{.1} = .5000$; $\bar{p}_{.2} = .3125$; $\bar{p}_{.3} = .1875$

B.8 a. $p_{11} = .075(.5000) = .0375$; $\bar{p}_{13} = .075(.1875) = .0141$;
$p_{22} = .925(.3125) = .2891$

b. $F_{11} = 800(.0375) = 30.00$; $F_{13} = 800(.0141) = 11.28$;
$F_{22} = 800(.3891) = 231.28$

B.9 $\bar{p}_{1.} = .8167$; $\bar{p}_{.1} = .1333$; $\bar{p}_{11} = .1089$; thus, $F_{11} = 300(.1089) = 32.67$.

B.10 For each cell of the table, the residual $f_{ij} - F_{ij}$ is computed. Each residual is then squared and, finally, divided by the expected frequency for that cell, F_{ij} to give the relative squared residual $(f_{ij} - F_{ij})^2/F_{ij}$. X^2 is then obtained by summing the relative squared residuals for all eight cells.

B.11 $r = 2$; $c = 4$; $X^2(.99; 3) = 11.34$

B.12 Since the P-value $= .0606 < .10 = \alpha$, conclude H_1, the type of insurance and frequency of physical exam are not statistically independent.

B.13

Frequency of Physical Exam	Estimated Probability Conditioned Upon Type of Insurance	
	Prepaid	Not Prepaid
Every Year or More	.550	.450
Every Two to Five Years	.300	.409
Every Six Years or More	.150	.141
	1.000	1.000

The probability of receiving very infrequent physical exams (i.e., every six years or less) is similar among employees with prepaid insurance and those with other types of insurance. However, employees with prepaid insurance are relatively more likely to receive a physical exam every year or more and relatively less likely to receive one every two to five years when compared to employees with insurance that is not prepaid.

B.14 contingency; contingency table

Problem 2

a. H_0: $p_{ij} = p_{i.}p_{.j}$ for all (i, j);
   H_1: $p_{ij} \neq p_{i.}p_{.j}$ for some (i, j)

b. $1 - \alpha = .99$; $(r - 1)(c - 1) = (3 - 1)(3 - 1) = 4$;
   $X^2(.99;4) = 13.28$. Thus the decision rule is:
   If $X^2 \leq 13.28$, conclude H_0;
   If $X^2 > 13.28$, conclude H_1.

c. The expected frequencies are computed as

$$F_{ij} = \overline{np_{ij}} = \overline{np_{i.}} \; \overline{p_{.j}} = n \frac{f_{i.}}{n} \frac{f_{.j}}{n} = \frac{f_{i.} \, f_{.j}}{n}$$

For example, $F_{11} = \dfrac{60(60)}{300} = 12$ and $F_{12} = \dfrac{60(150)}{300} = 30$

The remaining F_{ij}'s are computed in similar fashion, yielding $F_{13} = 18$, $F_{21} = 42$, $F_{22} = 105$, $F_{23} = 63$, $F_{31} = 6$, $F_{32} = 15$, and $F_{33} = 9$.

d. $X^2 = (4-12)^2/12 + (30-30)^2/30 + \ldots + (4-9)^2/9 = 58.095$

e. Since $X^2 = 58.095 > 13.28$, conclude H_1, the variables age class and education level are not statistically independent.

f. i.

Education Level	Estimated Probability Conditioned Upon Age Class		
	Under 35	35-Under 50	50 And Above
No College Degree	.067	.200	.289
Baccalaureate	.583	.767	.667
Advanced Degree	.350	.033	.044
Total	1.000	1.000	1.000

    ii. The conditional distribution of education level among
       middle-level managers in the "Under 35" age class differs
       markedly from the other two conditional distributions. It
       appears that a substantially higher proportion of managers in
       this age class have an advanced degree and a substantially
       lower proportion have no college degree, as compared to
       managers in the other two age classes.

C.1   a. bivariate multinomial population

     b. two univariate multinomial populations

     c. three univariate multinomial populations

C.2 $c = 3$; $r = 2$

C.3 p_{12}

C.4   a. f_{12} is the observed frequency in the first attribute class for
       the second population.

     b. n_2 is the sample size for the second population.

     c. n_T is the total number of observations in the study.

     d. $\bar{p}_1$ is the proportion of observations in the first attribute
       class among all the samples combined. (Equivalently, p_1 is the
       estimator for the proportion of the population in the first
       attribute class under alternative H_0.)

     e. f_{12} is the expected frequency in the first attribute class for
       the second population under alternative H_0.

C.5   a. $f_{12} = 80$; $f_{21} = 95$; $f_{31} = 20$

     b. $n_1 = 300$; $n_2 = 400$; $n_T = 700$

     c. $\bar{p}_1 = 265/700 = .379$; $\bar{p}_2 = 285/300 = .407$; $\bar{p}_3 = 150/700 = .214$

     d. $F_{12} = 400(.379) = 151.6$;
       $F_{21} = 300(.407) = 122.1$;
       $F_{31} = 300(.214) = 64.2$

Problem 3

a. H_0: $p_{11} = p_{12}$, $p_{21} = p_{22}$ and $p_{31} = p_{32}$;
   H_1: not all equalities in H_0 hold

b. $1 - \alpha = .90$; $(r - 1)(c - 1) = (3 - 1)(2 - 1) = 2$;
   $X^2(.90; 2) = 4.61$. Thus, the decision rule is:
   If $X^2 \leq 4.61$, conclude H_0;
   If $X^2 > 4.61$, conclude H_1.

c. $\bar{p}_1 = 1128/1300 = .86769$; $\bar{p}_2 = 47/1300 = .03615$; $\bar{p}_3 = 125/1300 =$
   $.03615$; thus, $F_{11} = 800(.86769) = 694.15$ and $F_{12} = 500(.86769) =$
   433.85. Similarly, $F_{21} = 28.92$, $F_{22} = 18.08$, $F_{31} = 76.92$, and F_{32}
   $= 48.08$.

d. $X^2 = (696-694.15)^2/694.15 + (28-28.92)^2/28.92 + . . .$
   $+ (49-48.08)^2/48.08 = .1175$

e. Since $X^2 = .1175 < 4.61$, conclude H_0, the distribution of
   housing units by type of structure is identical in the two
   markets.

f. No. Having concluded that $p_{11} = p_{12}$, $p_{21} = p_{22}$, and $p_{31} = p_{32}$, we
   would not anticipate that the estimated probabilities will
   differ sharply from one another. In fact, this closeness of the
   estimated probabilities resulted in a X^2 value sufficiently
   small that we were led to conclude H_0 in the first place.

18

SIMPLE LINEAR REGRESSION

A. SIMPLE LINEAR REGRESSION MODEL
(Text: Chapter 18. Sections 18.1 and 18.2)

Review of Basic Concepts

A.1 In a regression study, it was desired to predict the market value of owner-occupied homes in a community from knowledge of household income. In this study, X and Y will be used to denote, respectively, which of the two variables? X _____
Y _____

A.2 For the situation in A.1, is the relationship between X and Y more likely to be functional or statistical? _____
Explain. _____

A.3 For each of the following, indicate whether the relation between the variables is functional or statistical.

   a. Number of years of work experience and productivity of workers in a particular factory. _____

   b. Today's high temperature measured in Fahrenheit and today's high temperature measured in Celsius. _____

   c. Daily revenue from sales of $0.50 ice cream cones at a refreshment stand and daily number of cones sold.
   _____

A.4 A graphical representation of the clustering of points about a line or curve of statistical relationship is called a(n) _____ plot.

A.5 The simple linear regression model (18.3) in your Text assumes that the error terms, ϵ_i, are independent $N(0, \sigma^2)$. Explain what is meant by this assumption. _____
_____
_____

A.6 In the simple linear regression function, $E(Y) = \beta_0 + \beta_1 X$, the parameter ßo is the value of the regression function when $X =$ _____ and the parameter β_1 is the _____ of the regression function.

A.7 Suppose that, in a statistical study of the relation between employee wage rate (Y) and employee years of experience (X), it is known that ßo = \$2.82 and β_1 = \$1.35, with $\sigma^2 = 1.3$.

   a. What is the simple linear regression function for this situation? _____

   b. By how much does the mean wage rate increase for each additional year's experience? _____

   c. According to the regression function, what is the mean wage rate for employees with 12 years experience? _____

   d. What does the simple linear regression model (18.3) in your Text assume about the distribution of wage rate among all employees with 12 years experience? _____

   _____

A.8 In a study of the relationship between monthly advertising expenditures (X) and monthly sales (Y) of a manufacturing firm, a scatter plot showed that all the sample cases fell close to the curve of statistical relation. Does this prove that particular levels of advertising expenditure cause particular sales levels? _____ Explain. _____

»Problem 1

   A study has been undertaken to examine the relationship between scores on the mathematics section of a graduate school entrance exam (Y) and number of undergraduate mathematics courses completed (X). Following are observations from seven students who have recently taken the entrance exam.

	Number of	
Case	Mathematics	Mathematics
i	Courses	Score
	X_i	Y_i
1	6	40
2	6	36
3	4	30
4	7	42
5	9	53
6	3	27
7	5	35

a. Construct a scatter plot for these seven cases.

b. Does the assumption of a linear relationship between X and Y appear warranted on the basis of your scatter plot? _____ Explain. _____ ✓

c. Suppose the regression function here is $E(Y) = 15 + 4X$ and that $\sigma^2 = 9$. Plot the regression function on the graph in part (a.).✓

d. Construct a graph to show, in relation to the regression line, the general shape of the distribution of mathematics scores for students who have completed 8 undergraduate mathematics courses. ✓

B. POINT ESTIMATION OF β_0, β_1 AND $E\{Y\}$
(Text: Chapter 18, Sections 18.3 and 18.4)

Review of Basic Concepts

B.1 Identify the symbol which denotes the estimator for each of the following.

a. β_0 _____

b. β_1 _____

c. Regression Function _____

d. Mean response when $X = X_h$ _____

e. ϵ_i _____

B.2 Given a scatter plot constructed from the sample cases for a regression study and any straight line through the scatter plot, the sum of the squared deviations of Y observations from the line is denoted by _____ .

B.3 For the situation in B.2, if the slope (b_1) and intercept (b_0) of the straight line are obtained from (18.10a) and (18.10b) in your Text, the value of Q is a(n) _____ . Thus, when b_1 and b_0 are obtained from these formulas, they are called _____ _____ _____ of β_1 and β_0.

B.4 Explain the meaning of the following.

a. (X_i, Y_i) _____

b. $\hat{Y}_i$ _____

c. X_h _____

d. $\hat{Y}_h$ _____

B.5 Preliminary computation in a regression study with 10 cases yielded the following: $\Sigma X_i = 30$; $\Sigma Y_i = 460$; $\Sigma X_i^2 = 116$; $\Sigma X_i Y_i = 1742$. Using (18.10a) and (18.10b) in your Text, find the least squares estimates for β_1 and β_0.

a. $b_1 = $ _____

b. $bo = $ _____

B.6 For the situation in B.5, the estimated regression function is _____ .

B.7 The least squares estimates in a particular study are $b_0 = 10$ and $b_1 = 5$. The fourth case used in computing these estimates was ($X_4 = 2$, $Y_4 = 7$).

For this case, $Y_4 = $ _____ and is called the _____ value of $Y_4 = 7$. Also, $e_4 = $ _____ and is called the _____ for this case.

B.8 For the situation in B.7, the estimated mean response when $X_h = 5$ is denoted by _____ and can be computed to be _____ .

B.9 In least squares estimation for the simple linear regression model
(18.3) in your Text, a property of the residuals is that Σe_i =
_____. ✓

»Problem 2

A real estate firm has six offices of identical size in a
midwestern metropolitan area. It wishes to study the relationship
between the cost of heating an office in dollars per day (Y), and office
age in years (X). Records for a representative heating period are
presented in Table 18.1. The firm assumes there is a linear statistical
relationship between heating cost and office age and wishes to estimate
the regression function.

Table 18.1
Observations for Office Age (X_i) and Heating Cost (Y_i)
For Six Real Estate Offices
(See Problem 2)

Office i	(1) Office Age X_i	(2) Heating Cost Y_i	(3) $X_i Y_i$	(4) X^2	(5) Fitted $\hat{Y}_i$	(6) Residual e_i
1	1	2.0	2.0	1	____	____
2	3	3.0	9.0	9	____	____
3	4	2.5	____	____	____	____
4	2	2.0	____	____	____	____
5	1	2.0	____	____	____	____
6	7	3.5	____	____	____	____

Total: _____   _____   _____   _____

a. Complete the computations in columns (1) to (4) in Table 18.1. ✓

b. i. Use the method of least squares to obtain point estimates for
   β_1 and β_0. b_1 = _____ b_0 = _____

  ii. State the estimated regression function. _____ ✓

c. i. Construct a scatter plot of the six sample cases.

  ii. On the same graph, plot the estimated regression function. ✓

d. i. Determine $\hat{Y}_h$ when $X_h = 6$. _____

   ii. Interpret the value $\hat{Y}_h$ obtained in (i.) _____

_____ √

e. i. Using columns (5) and (6) in Table 18.1, compute the fitted values and residuals for the six cases.

   ii. Equation (18.18) in your Text states that the least squares residuals sum to zero. Do your computations in (i.) support this statement? _____ √

C. ANALYSIS OF VARIANCE APPROACH TO SIMPLE LINEAR REGRESSION
(Text: Chapter 18, Section 18.5)

Review of Basic Concepts

C.1 Identify the sum of squares symbol used to denote each of the following measures of variability.

a. Variability of the Y observations in the absence of knowledge concerning the independent variable X.

b. Reduction in variability of the Y observations associated with use of the regression relation.

c. Variability of the Y observations that remains as residual variation when the regression relation is considered. _____

C.2 SSTO measures variability of the Y observations around _____. Hence, SSTO is computed as $\Sigma($_____$)^2$.

C.3 SSE measures variability of the Y observations around _____. Hence, SSE is computed as $\Sigma($_____$)^2$.

C.4 If the residuals in a regression study have been computed, SSE can be determined by _____ each residual and _____ these values.

C.5 A sum of squares divided by its associated degrees of freedom is called a(n) _____ _____.

C.6 The simple linear regression model (18.3) in your Text assumes that the error terms have constant variance σ^2. An unbiased estimator of σ^2 is given by which of the following: SSTO, SSE, SSR, MSE, or MSR? _____

C.7 Results of a simple linear regression study with 25 cases include the following: SSTO = 6240 and SSE = 3105. In this situation SSR = _____.

C.8 For the situation in C.7, determine the degrees of freedom associated ith each sum of squares.

SSR _____ ;   SSE _____ ;   SSTO _____

C.9 For the situation in C.7, MSR = _____ and MSE = _____. Hence, an estimate of σ, the standard deviation of the error terms can be computed as _____.

C.10 In a regression study, SSTO = 624 and all cases fall exactly on the estimated regression line. Thus, SSE = _____ and SSR = _____ .✓

»Problem 3

a. Refer to the situation described in Problem 2. Complete the following table of computations for these R = 6 cases. ✓

i	X_i	Y_i	$\overline{Y}$	$\hat{Y}_i$	$(Y_i - \overline{Y})^2$	$(Y_i - \hat{Y}_i)^2$
1	1	2.0	2.5	2.00	.25	0
2	3	3.0	2.5	2.50	.25	.2500
3	4	2.5	2.5	2.75	_____	_____
4	2	2.0	2.5	2.25	_____	_____
5	1	2.0	2.5	2.00	_____	_____
6	7	3.5	2.5	3.50	_____	_____

Total _____ _____

b. Using the results in (a.) and your knowledge that regression and error sums of squares are additive, determine each of the following.

SSTO = _____ ; SSE = _____ ; SSR = _____ . ✓

c. Verify your results in (b.) by computing the sums of squares for this problem with the computational formulas (18.29), (18.30), and (18.31) in your Text. (Note: ΣX_i = 18.0, ΣY_i = 15.0, ΣX_i^2 = 80, ΣY_i^2 = 39.5, $\Sigma X_i Y_i$ = 51.5 and n = 6.)

i.   SSTO _____

ii. SSR _____

iii. SSE _____ √

d. Construct the ANOVA table for this regression study.

e. Obtain a point estimate of σ^2. _____ √

D. COEFFICIENTS OF SIMPLE DETERMINATION AND CORRELATION
(Text: Chapter 18, Sections 18.6 and 18.7)

Review of Basic Concepts

D.1 The coefficient of simple determination, r^2 can be no larger than _____. If the statistical relation is perfectly linear, then SSE = _____. Hence, SSR _____ SSTO and r^2 = SSR/SSTO = _____.

D.2 It is also true the r^2 can be no smaller than _____. If X is of no use in reducing the variability in Y, then SSR = _____ and, hence, r^2 = SSR/SSTO = _____.

D.3 Values of r^2 which are close to _____ suggest that the regression relation accounts for most of the variability in the dependent variable.

D.4 In a study of the relationship between age of child (X_i) and the number of hours of TV watched per week (Y_i), it is found that r^2 = .64. Explain what is meant by this value r^2 = .64. _____

_____

_____

D.5 For the situation in D.4, the estimated regression function is $\hat{Y} =$ 11.21 + .65X. Thus, the coefficient of simple correlation, which is denoted by _____ , is equal to _____ .

D.6 In a simple regression study, it is found that $b_1 = 0$. Thus, $r^2 =$ _____ and $r =$ _____ .

D.7 The estimated regression function in a study is $\hat{Y} = 46.7 - 3.2X$. Also, $r^2 = .36$ in this study. Thus, $r =$ _____ . √

»**Problem 4**

Problems 2 and 3 have examined the statistical relation between heating cost (Y_i) and age of office (X_i). It has been found that $\hat{Y} =$ 1.75 + .25X, SSTO = 2.0, SSR = 1.625, and SSE =.375.

a. Compute r^2 for this study. √

b. By what percent is the variability in heating cost reduced when office age is considered? _____

c. Compute r, the coefficient of simple correlation. √

Answers to Chapter Reviews and Problems

A.1 X denotes household income; Y denotes market value of home.

A.2 Statistical. It is very unlikely that the market value of a home is uniquely determined when the value of household income is specified. Rather, it is likely that market value is inherently variable for a given income level since not all households with the same income would choose to live in homes of exactly the same value. Further, other factors, such as location of the home within the community, would likely affect the market value of the home.

A.3 a. statistical

b. functional

c. functional

A.4 scatter

A.5 The assumption means that the error terms are normally distributed with expected value of 0 and constant variance σ^2, and that error terms for different observations are statistically independent.

A.6 0; slope

A.7 a. $E\{Y\} = 2.82 + 1.35X$

   b. $\beta_1 = \$1.35$

   c. $E\{Y\} = 2.82 + 1.35(12) = \$19.02$ when $X = 12$.

   d. Wage rates among employees with 12 years experience are normally distributed with mean 19.02 and variance 1.3 (i.e., Y's are independent $N(19.02, 1.3)$ when $X = 12$.

A.8 No. It may be that advertising causes sales levels. On the other hand, advertising levels may be set as a multiple of sales. Hence, the fact that the two variables are statistically related does not establish the direction of causation or even prove that a cause and effect relation exists between them.

Problem 1

   a. (See answer to part (c.) of this problem.)

   b. Yes; the scatter plot indicates that the cases follow a straight line and, hence, that the statistical relationship is linear, at least within the range of X values observed.

c.

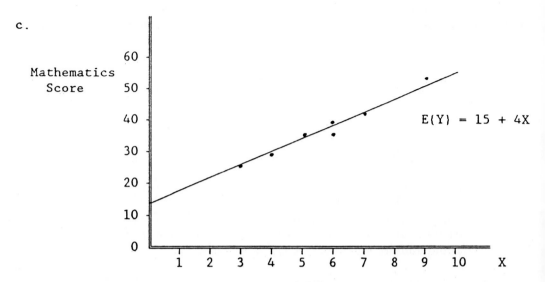

Figure 18.1

Number of Mathematics Courses

d.

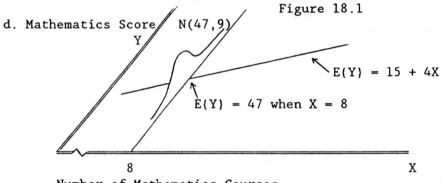

Number of Mathematics Courses

Figure 18.2

B.1 a. b_0

b. b_1

c. $\hat{Y} = b_0 + b_1X$

d. $\hat{Y}_h = b_0 + b_1X_h$

e. e_i

B.2 Q

B.3 minimum; least squares estimators

B.4 a. (X_i, Y_i) denotes the ith sample case. That is, X_i denotes the value of the independent variable and Y_i denotes the value of the dependent variable for the ith sample case.

b. $\hat{Y}_i$ denotes the value of the estimated regression function when $X = X_i$, the value of the independent variable for the ith sample case.

c. X_h denotes any specified level of the independent variable.

d. $\hat{Y}_h$ denotes the value of the estimated regression function when $X = X_h$, a specified level of the independent variable.

B.5 a. $b_1 = \dfrac{1742 - (30)(460)/10}{116 - (30)^2/10} = \dfrac{362}{26} = 13.9231$

b. $b_0 = (1/10)[460 - (13.9231)(30)] = 4.2307$

B.6 $\hat{Y} = 4.2307 + 13.9231X$

B.7 $\hat{Y}_4 = 10 + 5(2) = 20$; fitted; $e_4 = 7 - 20 = -13$; residual

B.8 $\hat{Y}_h = 10 + 5(5) = 35$ when $X_h = 5$

B.9 $\Sigma e_i = 0$

Problem 2

a.

i	(1) X_i	(2) Y_i	(3) $X_i Y_i$	(4) X_i^2
1	1	2.0	2.0	1
2	3	3.0	9.0	9
3	4	2.5	10.0	16
4	2	2.0	4.0	4
5	1	2.0	2.0	1
6	7	3.5	24.5	49
Total	$\Sigma X_i = 18$	$\Sigma Y_i = 15.0$	$\Sigma X_i Y_i = 51.5$	$\Sigma X^2 = 80$

$$b.i.b_1 = \frac{51.5 - (18)(15)/6}{80 - (18)^2/6} = .250; \quad b_0 = 1/6[15.0 - .250)(18) = 1.750$$

ii. $\hat{Y} = 1.75 + .25X$

c.i. and ii.

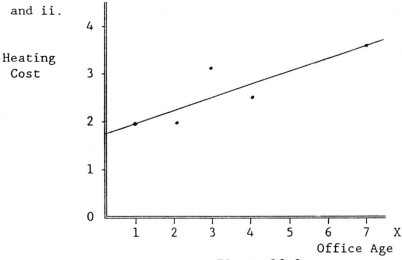

Figure 18.3

d. i. $\hat{Y}_h = 1.75 + .25(6) = 3.25$ when $X_h = 6$.

   ii. The estimated mean heating cost among a large number of six-
       year-old offices of this size is \$3.25 per day.

e. i.

i	(5) $\hat{Y}_i = 1.75 + .25X_i$	(6) $e_i = Y_i - \hat{Y}_i$
1	2.00	0
2	2.50	.50
3	2.75	-.25
4	2.25	-.25
5	2.00	0
6	3.50	0

   ii. Yes; $\Sigma e_i = 0 + .50 + (-.25) + (-.25) + 0 + 0 = 0$

C.1 a. SSTO

    b. SSR

    c. SSE

C.2 $\overline{Y}$ or the mean of the Y observations; SSTO = $\Sigma(Y_i - \overline{Y})^2$

C.3 $\hat{Y}_i$ or the fitted regression line; SSE = $\Sigma(Y_i - \hat{Y})^2$

C.4 squaring; summing

C.5 mean square

C.6 MSE

C.7 SSR = 6240 - 3105 = 3135

C.8 1; 25 - 2 = 23; 25 - 1 = 24

C.9 MSR = 3135/1 = 3135; MSE = 3105/23 = 135; $\sqrt{MSE}$ = $\sqrt{135}$ = 11.619

C.10 SSE = 0 ; SSR = 624

Problem 3

    a.

i	$(Y_i - \overline{Y})^2$	$(Y_i - \hat{Y}_i)^2$
1	.25	0
2	.25	.2500
3	.0-	.0625
4	.25	.0625
5	.25	0
6	1.00	0
Total	2.00	.3750

    b. SSTO = $\Sigma(Y_i - \overline{Y})^2$ = 2.00; SSE = $\Sigma(Y_i - \overline{Y})^2$ = .375;
       SSR = 2.00 - .375 = 1.625

    c. i.   SSTO = 39.5 - $(15)^2/6$ = 2.00

       ii.   SSR = $\dfrac{[51.5 - (18)(15)/6]^2}{80 - (18)^2/6}$ = $\dfrac{(6.5)^2}{26}$ = 1.625

       iii. SSE = 2.00 - 1.625 = .375

d.

Source of Variation	SS	df	MS
Regression	1.625	1	1.625M
Error	.375	4	.09375
Total	2.000	5	

e. MSE = .09375

D.1 1; SSE = 0 ; SSR = SSTO;, r^2 = SSR/SSTO = 1

D.2 0; SSR = 0 ; r^2 = SSR/SSTO = 0/SSTO = 0

D.3 1.0

D.4 The value r^2 = .64 means that the variability in hours of TV watched per week is reduced 64 percent when age of child is considered.

D.5 r; $+\sqrt{.64}$ = +.80, with the sign of b_1(i.e., b_1 = +.65) determining the sign of r.

D.6 0; 0

D.7 $\sqrt{.36}$ = -.60

Problem 4

a. r^2 = 1.625/2.0 = .8125

b. 81.25 percent

c. r = $+\sqrt{.8125}$ = +.9014

19

SIMPLE LINEAR REGRESSION - II

A. CONFIDENCE INTERVAL FOR MEAN RESPONSE
(Text: Chapter 19, Section 19.1)

Review of Basic Concepts

A.1 $\hat{Y}_h$ is a random variable with the sampling distribution described by (19.3) in your Text. Explain why $\hat{Y}_h$ is a random variable. _____

_____

_____

A.2 The mean of the sampling distribution of $\hat{Y}_h$, $E\{\hat{Y}_h\}$ = _____.

A.3 The variance of the sampling distribution of $\hat{Y}_h$ is denoted by ____.

A.4 Indicate whether each of the following statements about the sampling distribution of $\hat{Y}_h$ is true or false.

   a. "The mean of the sampling distribution of $\hat{Y}_h$ is the same for all levels of $X = X_h$." _____

   b. "The variance of the sampling distribution of $\hat{Y}_h$ is the same for all levels of $X = X_h$." _____

   c. "The functional form of the sampling distribution of $\hat{Y}_h$ is the same for all levels of $X = X_h$." _____

A.5 One component of $\sigma^2\{\hat{Y}_h\}$, the variance of the sampling distribution of $\hat{Y}_h$, is usually unknown. This component is _____. Its unbiased estimator is _____. The resulting estimated variance of $\hat{Y}_h$ is denoted by the symbol _____.

A.6 Indicate what effect each of the following has on the magnitude of $s^2\{\hat{Y}_h\}$.

   a. As the variability of the X_i's around their mean increases, the magnitude of $s^2\{\hat{Y}_h\}$ tends to _____.

b. As the sample size (i.e., number of cases) is increased, the magnitude of $s^2(\hat{Y}_h)$ tends to _____ .

c. As the difference between the specified level X_h and $\bar{X}$ increases, the magnitude of $s^2(\hat{Y}_h)$ tends to _____ .

d. As MSE decreases, the magnitude of $s^2(\hat{Y}_h)$ tends to _____ .

A.7 In a simple linear regression study with 20 cases, a 95% confidence interval for $E\{Y_h\}$ is desired. The appropriate value of the t distribution to use in constructing this confidence interval is _____ .

A.8 Results of a regression study with 5 cases were $\hat{Y} = -12.40 + 4.24X$ and $s^2(\hat{Y}_h) = 1.96$ when $X_h = 10$. Determine each of the following if a 90 percent confidence interval for $E\{Y_h\}$ when $X_h = 10$ is desired.

a. $\hat{Y}_h = $ _____

b. $s(\hat{Y}_h) = $ _____

c. $t(1 - \alpha/2;n - 2) = $ _____

A.9 For the situation in A.8, the appropriate confidence limits are _____ $\pm$ _____ . Hence, the 90 percent confidence interval for $E\{\hat{Y}_h\}$ when $X_h = 10$ is _____ $\leq \hat{Y}_h \leq$ _____ . $\checkmark$

»Problem 1

A regression study involving office age (X) and cost of heating office (Y) was introduced in Chapter 18, Problem 2. The six cases (X_i, Y_i) are presented in Table 18.1. Pertinent results for this study are $\hat{Y} = 1.75 + .25X$ and MSE $= .09375$.

a. Suppose a 90 percent confidence interval for mean heating costs among five year old offices of this size is desired.

   i. Using the observations for office age in Table 18.1, compute $\Sigma(X_i-X)^2$. _____

   ii. Compute $\hat{Y}_h$ when $X_h = 5$. _____

iii. Compute $s^2\{\hat{Y}_h\}$ when $X_h=5$. _____ ✓

iv. Obtain a 90 percent confidence interval for $E\{Y_h\}$ when $X_h = 5$. _____ ✓

b. Obtain a 90 percent confidence interval for $E\{Y_h\}$ when $X_h = 3$. _____ ✓

c. Is the confidence interval for $E\{Y_h\}$ when $X_h = 3$ wider or narrower than that obtained when $X_h = 5$? _____ Why? _____

B. PREDICTION OF NEW RESPONSE $Y_{h(new)}$
(Text: Chapter 19, Section 19.2)

Review of Basic Concepts

B.1 For each of the following, indicate whether interest centers on a mean response $E\{Y_h\}$ or new response $Y_{h(new)}$.

a. The personnel department of a large company uses an aptitude test to predict productivity of job applicants. Joe Daley has just completed a job application and taken the aptitude test, and the personnel manager wishes information about Joe's productivity. _____

b. For workers who have been out of high school for 10 years, which group tends to have higher salaries, those with 4 years of college or those with 6 years of college? _____

c. What is the average amount of life insurance for males having $50,000 annual incomes? _____

d. What will be the starting salary for Vicki Driver, a recent criminal justice graduate who has an undergraduate grade point average of 3.5? _____

B.2 When constructing an interval to predict a new response $Y_{h(new)}$ rather than to estimate a mean response $E\{Y_h\}$ for the same X_h, how do the degrees of freedom of the corresponding t values compare to one another? _____

B.3 When constructing a prediction interval for a new response $Y_{h(new)}$ when the regression parameters are unknown, the prediction limits must take into account two sources of variation. Identify these two sources of variation.

a. _____

b. _____

B.4 The appropriate estimated variance for a prediction interval for $Y_{h(new)}$ is denoted by _____.

B.5 In a simple linear regression study, the following results have been obtained: MSE = 620 and $s^2(\hat{Y}_h)$ = 11.5 when X_h = 85. Thus, the estimated variance for a prediction interval for a new response for which X_h = 85 is _____.

B.6 A researcher has developed a regression model to predict weekly food expenditures on the basis of family size. As part of his analysis, the researcher obtained a confidence interval for mean food expenditures among families of size four and a prediction interval for food expenditures of a particular family of size four. Which of these two intervals must necessarily be wider? _____ Explain. _____

_____ ✓

»Problem 2

Refer to Problem 1. Suppose two offices of this size have recently been acquired. One of these is five years old and the other is four years old. It is desired to predict daily heating costs at each of these two offices.

a. Construct a 90 percent prediction interval for daily heating costs for the five year old office. (Note: From Problem 1, $\hat{Y}_h$ = 3.0 and $s^2(\hat{Y}_h)$ = .0300 when X_h = 5, and MSE = .09375.)
_____ ✓

b. Construct a 90 percent prediction interval for daily heating costs for the four year old office. _____ ✓

c. i. Why is the 90 percent prediction interval for the five year old office wider than that for the four year old office?

_____

   ii. What generalization does this suggest? _____

   _____ √

C.1 INFERENCES CONCERNING β_1 AND β_0
(Text: Chapter 19, Sections 19.3 to 19.5)

Review of Basic Concepts

C.1 The least squares estimator b_1 is an unbiased estimator of the slope of the regression line. Therefore, $E\{b_1\}$ =

_____ .

C.2 The magnitude of $s^2\{b_1\}$, the estimated variance of b_1, tends to decrease as the magnitude of MSE _____ and as the variability of the X_i's _____ .

C.3 There is no linear relationship between X and Y when β_1 =

_____ .

C.4 Results of a simple regression study with 25 cases are as follows: $\hat{Y} = 65 - 5X$, MSE = 250 and $\Sigma(X_i - X)^2 = 1000$. If a 99 percent confidence interval for β_1 is desired, we require $s^2\{b_1\}$ =

_____ , $s\{b_1\}$ = _____ and $t(1 - \alpha/2; n - 2)$ = _____ . Hence, the confidence limits are _____ ± _____ .

C.5 For the situation in C.4, the appropriate decision rule to test whether or not $\beta_1 = 0$, with $\alpha = .01$, is _____

_____ .

C.6 For the situation in C.4 and C.5, the standardized test statistic $t^* =$ _____ , and $|t^*|$ = _____ . Thus, alternative _____ is concluded.

C.7 For the situation in C.4, suppose the 99 percent confidence interval for β_0 is $54 \leq \beta_0 \leq 76$. The 99 percent confidence interval for $E\{Y_h\}$ when $X_h = 0$ is then _____ . This is so for what reason? _____

_____

»**Problem 3**

Refer once again to the regression study involving the relationship between office age (X) and heating costs (Y). Recall that MSE = .09375, $\Sigma(X_i - X)^2 = 26$, X = 3, and $\hat{Y} = 1.75 + .25X$.

a. Conduct a test of whether or not $\beta_1 = 0$, with $\alpha = .10$. Specify the alternatives, determine t* and the appropriate decision rule and state the resulting decision. √

b. Construct a 90 percent confidence interval for β_1. _____ √

c. Based on the confidence interval in (b.), what can be said about the magnitude of increase in mean heating cost for each one year increase in office age? _____

_____

_____

»**Problem 4**

A regression study has been designed to examine the relation between the number of drums of an industrial solvent produced in a production run (X) and the total dollar cost of the production run (Y). Pertinent results for this study are $\hat{Y}=382.5 + 41.3X$, MSE = 31,600, $\Sigma(X_i - X)^2 = 55.0$, and n=15.

a. Conduct a test of whether or not $\beta_1 = 0$ using the standardized test statistic t*, with $\alpha = .05$. Specify the alternatives, determine t* and the appropriate decision rule and state the resulting decision. √

b. i. Explain the meaning of the parameter β_0 in the regression function for this study, assuming that the total cost for a production run consists of a fixed component, which does not depend on the number of drums produced in a run, and a variable component, which depends linearly on the number of drums produced in a run. _____

_____

_____

ii. Assuming the linear regression model is appropriate for X values as small as $X_h = 0$, construct a 95 percent confidence interval for β_0. _____

iii. Interpret the confidence interval for β_0 obtained in (ii.).

_____

_____ √

D. F TEST OF $\beta_1 = 0$
(Text: Chapter 19, Section 19.6)

Review of Basic Concepts

D.1 E(MSE) is always equal to σ^2. E(MSR) is also equal to σ^2 if
_____ equals zero.

D.2 The test statistic F* which is used to examine the alternatives H_0:
$\beta_1 = 0$ and H_1: $\beta_1 \neq 0$ is computed as the ratio _____/_____ .

D.3 For the situation in D.2, if $\beta_1 = 0$, the numerator and denominator
of F* have the same expected value, namely, _____ . If
$\beta_1 \neq 0$, the expected value of MSR is _____ than σ^2.
Hence, alternative H_0: $\beta_1 = 0$ should be selected for relatively
_____ values of F* and alternative H_1: $\beta_1 \neq 0$ should be
selected for relatively _____ values of F*.

D.4 Using Table C-4 in your Text, determine each of the following F
distribution percentiles.

a. F(.95; 1, 8) = _____

b. F(.95; 1, 18) = _____

c. F(.99; 1, 2) = _____

D.5 What is the decision rule in an F test of $\beta_1 = 0$ for a simple
regression study with 20 observations, given $\alpha = .05$? _____

_____ √

»Problem 5

A regression study involving n = 30 convenience grocery stores was undertaken to examine the relationship between X = monthly newspaper advertising expenditures and Y = number of customers shopping at a store. The ANOVA table for this study is as follows.

Source of Variation	SS	df	MS
Regression	2350	1	2350
Error	60480	28	2160
Total	62830	29	

a. State the alternatives for testing whether or not X and Y are linearly related. _____

b. Construct the appropriate decision rule to conduct this F test, with α = .01. _____

_____

c o Compute F*. _____

d. On the basis of this F test, can we conclude that X and y are linearly related? _____ Why? _____

_____

e. Would the conclusion in (d.) have differed had we conducted a test of whether or not β_1 = 0 with a test statistic based on the t distribution, with α=.01? _____. Explain. _____

_____

E. OTHER CONSIDERATIONS IN REGRESSION ANALYSIS
(Text: Chapter 19, Sections 19.7 and 19.8)

Review of Basic Concepts

E.1 The simple linear regression model (19.1) in your Text assumes that the regression function is _____ in form, and that the error terms, ϵ_i, are _____ distributed with _____ variance and are statistically _____ from one another.

E.2 In a residual plot, the residuals are plotted against the _____
_____ .

E.3 Which assumption of the simple linear regression model appears to have been violated based on evidence in each of the following residual plots?

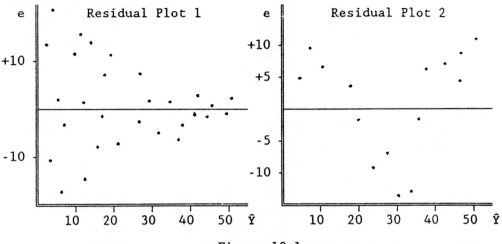

Figure 19.1

a. Residual Plot 1. _____

b. Residual Plot 2. _____

E.4 Identify two remedies that are available in instances where a scatter plot or residual plot gives evidence that the regression relation is not linear. _____

E.5 One check for normality of the error terms employs standardized residuals obtained by dividing each residual e_i by _____ .

E.6 If the cases for a regression study have been obtained in a time sequence, a plot of the residuals e_i against time will help us determine whether the error terms ϵ_i violate the assumption of _____ . If so, the error terms are said to be
_____ .

E.7 A regression study was designed to examine the relationship between number of units of output in a production run (X) and total cost of the production run (Y). Twenty production runs were used in the study, ranging in size from 1500 to 4500 units. The estimated regression function was $\hat{Y} = 383 + .24X$. Would it be wise, in this case, to conclude that the fixed cost of setting up a production run is near $b_0 = \$383$? _____ Explain. _____

E.8 For the situation in E.7, under what condition would X be viewed as random? _____

E.9 For the situation in E.7, suppose a new plant layout is instituted in the factory. Explain why the results of the regression study may no longer be appropriate. _____

E.10 Based on an F test of $\beta_1 = 0$, it was concluded from a study of a sample of colleges that the number of men who try out for a collegiate football team in a season (X) is related to total seasonal attendance (Y). What does this imply about the cause-effect relation between X and Y? _____
Explain. _____

E.11 For the situation in E.10, can you suggest an intervening variable which might explain the regression relation between X and Y? _____ Explain. _____

»**Problem 6**

A marketing analyst for a national chain of retail bookstores has used the simple linear regression model to relate Y = monthly sales (in $ thousands) to X = store frontage (in meters). The fitted values and residuals for the sixteen stores included in this study are presented in Table 19.1.

Table 19. 1
FITTED VALUES AND RESIDUALS FOR
SIXTEEN RETAIL BOOKSTORES
(See Problem 6)

Store i	Fitted Value $\hat{Y}_i$	Residual e_i	Store i	Fitted Value $\hat{Y}_i$	Residual e_i
1	31.1	-4.1	9	23.4	-4.7
2	15.1	-2.7	10	29.9	-3.7
3	22.2	-4.2	11	37.1	3.7
4	3.0	6.2	12	9.2	2.3
5	28.5	-5.0	13	7.4	4.2
6	33.3	2.9	14	18.5	-3.3
7	40.8	3.1	15	11.1	-1.2
8	42.8	5.6	16	32.1	0.9

a. Construct a plot of the residuals against the fitted values. √

b. What should be the appearance of this residual plot if the linear regression function is apt? _____

_____

c. Based on the residual plot obtained in (a.), does the linear regression function appear apt in this situation? _____
Explain. _____

_____ √

»Problem 7

The director of a regional ski association is studying the relation between annual number of ski related injuries at a ski area which require first aid attention (Y) and total number of skier days at the ski area (X). A regression study is being conducted with data from twenty ski areas in the region. The residuals for this study are presented in Table 19.2. Previous analysis of the residuals indicates that a linear fit with constant variance is appropriate. The director now wishes to check for normality of the error terms.

Table 19.2
RESIDUALS FOR STUDY OF SKI RELATED INJURIES
(See Problem 7)

i	Residual e_i	Standardized Residual e'_i	i	Residual e_i	Standardized Residual e'_i
1	-2.04	-.251	11	-3.61	-.444
2	-19.80	-2.435	12	-3.16	-.389
3	1.73	.213	13	9.87	1.214
4	-4.09	-.503	14	-7.32	-.900
5	3.12	.384	IS	9.65	1.187
6	14.89	1.831	16	3.49	_____
7	-4.59	-.564	17	-4.65	_____
8	7.29	.897	18	5.28	_____
9	-6.27	-.771	19	2.07	_____
10	-7.80	-.959	20	5.94	_____

a. Compute the standardized residuals for the last five cases in Table 19.2. (Note: Σe^2 = SSE = 1190.123. Hence, MSE = 1190.123/18 = 66.118. √

b. i. Using the 5th, 25th, 50th, 75th and 95th percentiles of the relevant t distribution, complete the following table of observed and expected frequencies for the standardized residuals.

| | | Frequency | |
Standardized Residual e'_i		Observed f	Expected F
under -1.734		_____	_____
_____		_____	_____
_____		_____	_____
_____		_____	_____
_____		_____	_____
Total		20.0	20.0

    ii. Is there evidence from these frequencies that the Y distributions are not normal? _____ Explain.

_____ √

Answers to Chapter Reviews and Problems

A.1 $\hat{Y}_h$ is a random variable since, for a specified level X_h the value of $\hat{Y}_h$ will vary from sample to sample. This is so because repeated samples will lead to different values for b_0 and b_1. As b_0 and b_1 vary, the value of $\hat{Y}_h = b_0 + b_1 X_h$ will similarly vary.

A.2 $E\{\hat{Y}_h\} = E\{\hat{Y}_h\} = \beta_0 + \beta_1 X_h$

A.3 $\sigma^2\{\hat{Y}_h\}$

A.4 a. False, provided $\beta_1 \neq 0$ b. False   c. True

A.5 σ^2; MSE; $s^2\{\hat{Y}_h\}$)

A.6 a. decrease   b. decrease   c. increase   d. increase

A.7 $t(.975; 18) = 2.101$

A.8 a. $\hat{Y}_h = -12.40 + 4.24(10) = 30.0$

    b. $s\{\hat{Y}_h\} = \sqrt{1.96} = 1.40$

    c. $t(.95; 3) = 2.353$

A.9 $30.0 \pm 2.353(1.40)$; $26.706 \leq E\{Y_h\} \leq 33.294$

Problem 1

a. i. From Table 18.1, $\bar{X} = 18/6 = 3$. Hence, $\Sigma(X_i - \bar{X})^2 =$
    $(1-3)^2 + (3-3)^2 + (4-3)^2 + (2-3)^2 + (1-3)^2 + (7-3)^2 = 26$

   ii. $\hat{Y}_h = 1.75 + .25(5) = 3.0$ when $X_h = 5$

  iii. $s^2\{\hat{Y}_h\} = .009375[1/6 + (5-3)^2/26] = .0300$ when $X_h = 5$

   iv. $s\{\hat{Y}_h\} = \sqrt{.0300} = .1732$; $t(.95; 4) = 2.132$. The confidence
       limits are $3.0 \pm 2.132(.1732)$. Thus, the 90 percent
       confidence interval for $E\{Y_h\}$ when $X_h = 5$ is $2.631 \le E\{Y_h\} \le$
       3.369.

b. $\hat{Y}_h = 1.75+.25(3) = 2.5$ when $X_h = 3$; $s^2\{\hat{Y}_h\} = .09375[1/6 +$

   $(3-3)^2/26]$ $.0156$; $s\{\hat{Y}_h\} = \sqrt{.0156} = .1249$; $t = (.95; 4) = 2.132$.
   The confidence limits are $2.5 \pm 2.132(.1249)$. Thus, the 90
   percent confidence interval for $E\{Y_h\}$ when $X_h = 3$ is $2.234 \le$
   $E\{Y_h\} \le 2.766$.

c. Narrower. Since the magnitude of $(X_h - \bar{X})^2 = (X_h - 3)$ is less
   when $X_h = 3$ than when $X_h = 5$, $s^2\{\hat{Y}_h\}$ is also smaller when $X_h = 3$
   than when $X_h = 5$. This, in turn, reduces the width of the
   confidence interval.

B.1 a. new response

   b. mean response

   c. mean response

   d. new response

B.2 The degrees of freedom are the same (i.e., $n - 2$).

B.3 a. The inherent variability in the probability distribution of Y,
       represented by σ^2.

   b. The uncertainty about the mean response at $X = X_h$, represented
      by $\sigma^2\{\hat{Y}_h\}$.

B.4 $s^2\{Y_{h(new)}\}$

B.5 $s^2(Y_{h(new)}) = 620 + 11.5 = 631.5$

B.6 The prediction interval must be wider. The terms in both intervals are identical except for the estimated standard deviations. Since $s(Y_{h(new)})$ is always larger than $s(\hat{Y}_h)$ when $X = X_h$, the prediction interval for a new response must be wider than the confidence interval for the mean for a specified level of X_h.

Problem 2

a. $s^2(Y_{h(new)}) = .09375 + .0300 = .12375$;
   $s^2(Y_{h(new)}) = \sqrt{.12375} = .3518$;
   $t(.95; 4) = 2.132$. The prediction limits are $3.0 \pm 2.1(.3518)$.
   Thus, the 90 percent prediction interval for $Y_{h(new)}$ when $X_h = 5$ is $2.250 \leq Y_{h(new)} \leq 3.750$.

b. $\hat{Y}_h = 1.75 + .25(4) = 2.75$ when $X_h = 4$;
   $s^2(Y_{h(new)}) = .09375[1 + 1/6 + (4-3)^2/26] + .11298$;
   $s^2(Y_{h(new)}) = \sqrt{.11298} = .3361$
   The prediction limits are $2.75 \pm 2.132(.3361)$. Thus, the 90% prediction interval for $Y_{h(new)}$ when $X_h = 4$ is $2.033 \leq Y_{h(new)} \leq 3,467$.

c. i. The prediction interval for the five year old office is wider because $(X_h - \bar{X})^2 = (X_h - 3)^2$ is larger and, hence $s^2(Y_{h(new)})$ is larger when $X_h = 5$ than when $X_h = 4$.

   ii. In general, the farther the specified level X_h deviates from $\bar{X}$ in either direction, the greater will tend to be $s^2(Y_{h(new)})$ and, hence, the wider will be a prediction interval for a given confidence coefficient.

C.1 β_1

C.2 decreases; increases

C.3 0

C.4 $s^2(b_1) = 250/1000 = .25$; $s(b_1) = .50$; $t(-995;23) = 2.807$;
   $-5 \pm 2.807(.50)$

C.5 If $|t*| \leq 2.807$, conclude $H_0(\beta_1 = 0)$;
   If $|t*| > 2.807$, conclude $H_1(\beta_1 \neq 0)$

C.6 $t* = -5/.50 = -10.0$; $|t*| = 10.0$; $H_1(\beta_1 \neq 0)$

C.7 $54 \leq E\{Y_h\} \leq 76$ when $X_h = 0$. The confidence intervals are the same since $E(Y_h) = \beta_0 + \beta_1 X_h = \beta_0 + \beta_1(0) = \beta_0$ when $X_h = 0$.

Problem 3

a. Alternatives: H_0: $\beta_1 = 0$; H_1: $\beta_1 \neq 0$

            t*: $s^2\{b_1\} = .09375/26 = .0036$;

                 $s\{b_1\} = \sqrt{.0036} = .06$;
                 Thus, $t* = 25/.06 = 4.167$.

    Decision Rule: $t(.95; 4) = 2.132$; thus, the decision rule is:
                 If $|t*| \leq 2.132$, conclude $H_0(\beta_1 = 0)$;
                 If $|t*| > 2.132$, conclude $H_1(\beta_1 \neq 0)$

        Decision: Since $|t*| = 4.167 > 2.132$, conclude $H_1(\beta_1 \neq 0)$, there is a linear relation between office age (X and heating cost (Y).

b. $s\{b_1\} = .06$; $t(95; 4) = 2.132$. The confidence limits are .25 ± 2.132(.06). Thus, the 90 percent confidence interval for β_1 is $.1221 \leq \beta_1 \leq .3779$.

c. With 90 percent confidence, we can state that mean daily heating cost increases between \$.1221 and \$.3779 for each one year increase in office age.

Problem 4

a. Alternatives: H_0: $\beta_1 = 0$; H_1: $\beta_1 \neq 0$

            t*: $s^2\{b_1\} = 31600/18370 = 1.7202$;
                $s\{b_1\} = \sqrt{1.7202} = 1.312$;
                Thus, $t* = 41.3/1.312 = 31.479$.

Decision Rule: t(.975; 13) = 2.160; thus, the decision rule is:
If $|t*| \leq 2.160$, conclude $H_0(\beta_1 = 0)$;
If $|t*| > 2.160$, conclude $H_1(\beta_1 \neq 0)$

Decision: Since $|t*| = 31.479 > 2.160$, conclude $H_1(\beta_1 \neq 0)$, there there is a linear relation between size of production run (X) and total production cost (Y).

b. i. β_0 is the intercept of the regression function. For this study, β_0 can be interpreted as the fixed cost associated with a production run.

ii. $b_0 = 382.5$;
$s^2(b_0) = 31600[1/15 + (55)^2/18370] = 7310.3$;

$s(b_0) = \sqrt{7310.3} = 85.50$;
t(.975; 13) = 2.160
The confidence limits are 382.5 ± 2.160(85.50). Thus, the 95% confidence interval for β_0 is $197.82 \leq \beta_0 \leq 567.18$.

iii. With 95% confidence, we can state that the fixed cost associated with a production run is between $197.82 and $567.18.

D.1 β_1

D.2 MSR/MSE

D.3 σ^2; larger; small; large

D.4 a. F(.95; 1, 8) = 5.32

b. F(.95; 1, 18) = 4.41

c. F(.99; 1, 25) = 7.77

D.5 F(.95;1, 18) = 4.41. Thus, the decision rule is:
If $F* \leq 4.41$, conclude H_0; if $F* > 4.41$, conclude H_1.

Problem 5

a. H_0: $\beta_1 = 0$; H_1: $\beta_1 \neq 0$

b. $F(.99; 1, 28) = 7.64$. Thus, the decision rule is: If $F* < \le .64$, conclude $H_0(\beta_1 = 0)$; if $F* > 7.64$, conclude $H_1(\beta_1 \ne 0)$.

c. $F* = 2350/2, 160 = 1.088$

d. No. Since $F* = 1.088 < 7.64$, conclude $H_0(\beta_1 = 0)$, that monthly newspaper advertising expenditures (X) and number of customers (Y) are not linearly related.

e. No; the same conclusion would have been reached. The t and F tests of whether or not $\beta_1 = 0$ will always result in the same conclusion when α is controlled at the same level in each test.

E.1 linear; normally; constant; independent

E.2 fitted values

E.3 a. Variance of the error terms does not appear to be constant.

b. Regression function does not appear to be linear.

E.4 First, the regression relation can often be linearized by employing a logarithmic or other transformation of the independent variable. Second, regression models other than the simple linear model (19.1) in your Text can be adopted.

E.5 $\sqrt{MSE}$

E.6 statistical independence, autocorrelated

E.7 No. The linear regression model, which was fit to X values only in the range $1500 \le X \le 4500$, may not be appropriate when $X_h = 0$. (Of course, if there is knowledge that the regression function is linear in the entire range $0 \le X \le 4500$ in this application, then b_0 can be interpreted as a reasonable estimate for the fixed cost component.)

E.8 X would be viewed as random if the size of a production run could not be controlled. This would be the case, for example, if the production run is determined by the size of a customer's (random) order or by a current inventory level which is dependent upon uncertain sales volume since the last production run.

E.9 Application of results of regression analysis to the future requires that the relevant causal conditions operating in the specific application do not change. In this situation, a new plant layout may well alter the cost structure of the production process.

E.10 Nothing. The existence of a regression relation between X and Y does not imply a causal relation, although there may well be one.

E.11 Size of College. Larger schools tend to have more students who try out for the football team and also tend to have larger attendances at football games.

Problem 6

a.

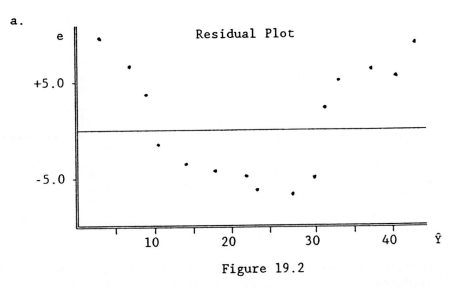

Figure 19.2

b. If the regression function is linear, the residuals will tend to scatter around the zero line at random, with no systematic pattern apparent.

c. No. In this residual plot, the residuals systematically lie above, then below, then again above the zero line as $\hat{Y}$ increases. A curvilinear regression function appears to be called for.

Problem 7

a. $e'_i = e_i / \sqrt{66.118} = e_i / 8.13$

i	Residual e_i	Standardized Residual e'_i
16	3.49	.429
17	-4.65	-.572
18	5.28	.649
19	2.07	.255
20	5.94	.731

b. i.

Standardized Residual e'			Observed f	Expected F
		under -1.734	1	1.0
-1.734	-	under -.688	3	4.0
-.688	-	under 0.0	6	5.0
0.0	-	under .688	5	5.0
.688	-	under 1.734	4	4.0
1.734	and	above	1	1.0
		Total	20.0	20.0

ii. A comparison of the observed and expected frequencies for the
standardized residuals reveals no evidence of serious departures
from normality. Indeed, the observed frequencies conform quite
well to those expected under the assumption that the error terms
(or, equivalently, the Y distributions) are normal.

A. MULTIPLE REGRESSION MODELS, CALCULATIONAL RESULTS AND RESIDUAL ANALYSIS

(Text: Chapter 20, Sections 20.1 to 20.3)

Review of Basic Concepts

A.1 Suppose a regression function is known to be $E(Y) = ß_0 + ß_1X_1 + ß_2X_2 = 14.5 + 2.0X_1 + 12.5X_2$. Interpret each of the $ß_i$ parameters for this situation.

a. $ß_0 = 14.5$ _____

b. $ß_1 = 2.0$ _____

c. $ß_2 = 12.5$ _____

A.2 For the situation in A.1, if the regression function is shown graphically, the resulting response surface is a(n) _____.

A.3 In the multiple regression model in p - 1 = 4 independent variables, explain what X_{i3} denotes. _____.

A.4 In a multiple regression model in p - 1 = 4 independent variables, identify the degrees of freedom associated with each sum of squares, given n = 50 cases.

a. SSR _____

b. SSE _____

c. SSTO _____

A.5 The ANOVA table for a multiple regression study with 100 cases and two independent variables is as follows. Complete this ANOVA table.

Source	SS	df	MS
Regression	46,488	_____	_____
Error	43,912	_____	_____
Total	90,400	_____	_____

A.6 For the situation in A.5, the least squares estimates β_0, β_1, and β_2 are 642.25, -14.60, and 1.07, respectively. The estimated regression function is _____ .

A.7 Information about the strength of linear relationship between pairs of variables in multiple regression analysis can be found in the _____ matrix.

A.8 The coefficient of simple correlation between the dependent variable Y and independent variable X_3 is denoted by _____ .

A.9 In model (20.1) in your Text, a point estimator for the variance of the Y distribution for a given X_1 and X_2 is _____ . For the situation in A.5, the appropriate point estimate is _____ .

A.10 The coefficient of multiple determination, R^2 can take on values between _____ and _____ . If all Y observations in a study fall exactly on the estimated response surface, R^2 = _____ .

A.11 For the situation in A.5, R^2 = _____ . If a third independent variable were added to this model, could R^2 decrease? _____ Explain. _____

_____

A.12 R^2 does not recognize the number of independent variables in the regression model. A modified measure that does recognize thenumber of independent variables is called the _____ coefficient of multiple determination, denoted by the symbol _____ .

A.13 The adjusted coefficient of multiple determination, defined in (20.12) in your Text, can decrease when an additional independent variable is added to a regression model. This is so because the degrees of freedom associated with SSE (i.e., n-p) is reduced by _____ with the added variable. Hence, R_a^2 will decrease unless the reduction in n - p is offset by a sufficient decrease in

_____

A.14 For the situation in A.5, R_a^2 = _____

A.15 For a regression analysis involving three independent variables, the error sums of squares for the regression models including and excluding X_3 are, respectively, SSE(X_1, X_2, X_3) = 1050 and SSE(X_1, X_2) = 1200. Thus, the coefficient of partial determination between the dependent variable and X_3 given that X_1 and X_2 are already in the model, which is denoted by _____, is equal to _____.

A.16 For the situation in A.15, interpret the value of the coefficient of partial determination. _____

_____

»Problem 1

A study involving toy safety has been undertaken to examine the relation between Y = strength (in maximum pounds of pressure per square centimeter) of a particle type of inflatable toy, X_1 = thickness of the material used in the toy (in mils) and X_2 = chemical density of the material used in the toy (measured as an index where $0.0 \leq X_2 \leq 5.0$. Observations on one toy of each of ten different brands appear in Table 20.1a. Computer output for a regression analysis of these data, using regression model (20.1) in your Text, is contained in Table 20.1b.

Table 20.1

REGRESSION RESULTS FOR STUDY OF TOY SAFETY
(See Problem 1)

(a)
Observations for Ten Brands

Brand i	Thickness of Material X_{i1}	Density of Material X_{i2}	Strength of Toy Y_i
1	1	2.8	20
2	1	3.8	32
3	4	2.9	42
4	4	3.8	66
5	6	3.8	93
6	1	3.0	15
7	2	3.2	23
8	3	3.1	37
9	4	3.6	49
10	4	4.0	83

(b)
Selected Computer Output for Regression Analysis

Variable	Reg. Coef.	Std. Dev.	T Stat.
Constant	-74.189	22.306	-3.33
X_1	10.626	1.872	5.68
X_2	25.973	7.154	3.63

Analysis of Variance Table

Source	SS	df	MS
Regression	5958.4	2	2979.20
Residual	487.6	7	69.66
Total	6446.0	9	

Variable	Specified Level
X_1	4.0
X_2	3.4

Y Estimate = 56.62   Std. Dev. Mean Response = 3.236
                          Std. Dev. New Response = 8.951

a. State the response function being estimated in this study.

_____

b. i. Give the estimated regression function. _____

   ii. Interpret the meaning of b_2, the estimated regression
       parameter associated with X_2; density of material.

       _____ ✓

c. Obtain and interpret the value of R^2 for this study. _____

_____

d. A regression analysis of Y on X_1 alone was previously performed,
   with the results $\hat{Y} = 4.231 + 13.923X_1$ and $SSE(X_1) = 1405.8$.
   Obtain and interpret the value of $r^2_{Y2.1}$ for this study.

   _____

   _____ ✓

B. STATISTICAL INFERENCES WITH MULTIPLE REGRESSION MODELS

(Text: Chapter 20, Sections 20.4 to 20.7)

»Problem 2

The Administrative Director for an association of city and county
zoos is studying the relation between y = annual number of visitors to a
zoo (in thousands), X_1 = total number of exhibits at the zoo, and X_2 =
age of zoo (in years). Observations for twenty zoos in a recent year
are presented in Table 20.2a. Computer output for a regression analysis
of these data using regression model (20.1) in your Text, is contained
in Table 20.2b.

a. Assuming regression model (20.1 in your Text is apt, conduct a
   test of whether or not $\beta_1 = \beta_2 = 0$, using $\alpha = .01$. Specify the
   alternatives, determine F* and the decision rule and state the
   resulting decision. ✓

b. Conduct a test of whether or not $\beta_1 = 0$, using $\alpha = .01$. Specify
   the alternatives, determine t* and the decision rule and state
   the resulting decision. ✓

c. Conduct a test of whether or not $\beta_2 = 0$, using $\alpha = .01$. Specify the alternatives, determine t* and the decision rule and state the resulting decision. √

Table 20.2

REGRESSION RESULTS FOR STUDY OF ANNUAL
NUMBER OF ZOO VISITORS
(See Problem 2)

(a)
Observations for Twenty Zoos

Zoo i	Number of Exhibits x_{i1},	Age of Zoo x_{i2},	Number of Visitors Y_i
1	50	31	145
2	36	3	190
3	130	26	255
4	228	32	1,280
5	193	56	800
6	138	50	385
7	67	1	120
8	162	13	520
9	178	14	555
10	85	59	225
11	117	55	270
12	130	23	350
13	96	20	320
14	214	49	965
15	119	34	230
16	41	33	105
17	105	18	200
18	160	61	425
19	153	41	370
20	94	11	105

(b)
Selected Computer Output for Regression AnAlysis

Variable	Reg. Coef.	Std. Dev.	T Stat.
Constant	-214.180	92.838	-2.31
X_1	5.008	.692	7.24
X_2	-.638	2.024	-.32

Analysis of Variance Table

Source	SS	df	MS
Regression	1396674	2	698337
Residual	403340	17	23726
Total	1800014	19	

»Problem 3

Refer to Problem 1. The data and computer output for this regression analysis are contained in Table 20.1. The F test of $\beta_1 = \beta_2 = 0$ with $\alpha = .05$ for this situation leads to conclusion H_1, that there is a regression relation between strength of the toy and the independent variables thickness of material (X_1) and density of material (X_2). Further tests with $\alpha = .05$ lead to the conclusions that $\beta_1 \neq 0$ and $\beta_2 \neq 0$. The analyst for this study now wishes to use the regression results in Table 20.1 to construct several confidence and prediction intervals. (Note: Estimated standard deviations for this problem appear in Table 20.1b).

a. i. Construct a 95 percent confidence interval for $\beta 1$.

_____

   ii. Construct a 95 percent confidence interval for β_2.

_____

   iii. Interpret these two confidence intervals. _____

_____

_____ √

b. i. Construct a 98 percent confidence interval for the mean strength of brands which are 4 mils thick with a density of 3.4. _____ √

   ii. Construct a 98 percent confidence interval for the strength of a single toy of a new brand which is 4 mils thick with a density of 3.4. _____

iii. Are the relative widths of these two intervals consistent with the relationship between confidence intervals for the mean response and prediction intervals for a new response in simple linear regression? _____ Explain. _____

_____ √

C. INDICATOR VARIABLES
(Text: Chapter 20, Section 20.8

C.1 The qualitative independent variable employee status (full-time versus part-time) is to be represented in a regression analysis with the 0, 1 indicator variable X_4, with "full-time" as the reference class. Thus, $X_4 = 0$ if employee status is _____ and $X_4 = 1$ if employee status is _____

C.2 Indicator variables are also called _____ variables.

C.3 For the regression model $E(Y) = \beta_0 + \beta_1 X_1 + \beta_2 X_2$, where X_2 is an indicator for a qualitative variable with two classes, there are two response functions. These functions have the same _____ but different _____ .

C.4 Suppose a response function is known to be $E(Y) = \beta_0 + \beta_1 X_1 + \beta_2 X_2 = 35 + 6X_1 + 15X_2$, where X_2 is an indicator for sex, coded $X_2 = 1$ if male and $X_2 = 0$ if female. Therefore, the response function for females is _____ and the response function for males is _____ .

C.5 For the situation in C.4, since $\beta_2 = 15$ is positive, the mean response for males for a given level X_{h1} is always _____ than the mean response for females, for the same level X_{h1}.

C.6 A regression study included a qualitative variable, type of residence, classified as either (1) private home, (2) small apartment building, (3) large apartment complex, or (4) condominium. In this study, a total of _____ indicator variables are required in the regression model.

C.7 For the situation in C.6, define an appropriate set of indicator variables. _____

_____ √

»Problem 4

An analyst for an auto parts manufacturer had casually observed that four wheel drive vehicles seem relatively more popular in snowbelt markets than in sunbelt markets, and in sparsely populated markets than in densely populated markets. To formally evaluate these observations, a multiple regression model was developed where Y = percent of privately owned vehicles in a market that are four wheel drive, X_1 = 1 if a snowbelt market and X_1 = 0 otherwise, and X_2 = population density (measured as hundreds of persons per square mile) of a market. Using regression model (20.1) and data from 68 markets (data not shown), the analyst obtained the computer output contained in Table 20.3.

Table 20.3

SELECTED COMPUTER OUTPUT FOR STUDY OF
FOUR WHEEL DRIVE VEHICLES
(See Problem 4)

Variable	Reg. Coef.	Std. Dv.	T Stat.
Constant	10.538	.431	24.45
X_1	~ 4.878	.391	12.48
X_1	-.588	.053	-11.09

Analysis of Variance Table

Source	SS	df	MS
Regression	757.20	2	378.600
Residual	162.75	65	2.504
Total	919.95	67	

a. Assuming regression model (20.1) in your Text is apt, conduct a test to determine whether or not there is a regression relation between the percent of privately owned vehicles in a market that are four wheel drive and this set of independent variables, using α = .01. Specify the alternatives, determine the value of the test statistic and the decision rule and state the resulting decision. ✓

b. Conduct a test to determine whether or not the indicator variable for snowbelt versus sunbelt markets can be dropped from the model which already contains the variable population density, using $\alpha = 01$. Specify the alternatives, determine the value of the test statistic and the decision rule and state the resulting decision. ✓

c. i. Identify the fitted (i.e., estimated) regression model for this study. _____

   ii. Interpret the coefficient associated with each independent variable in this fitted model.

X_1: _____

X_2: _____ ✓

D. MODELING CURVILINEAR RELATIONSHIPS
(Text: Chapter 20, Section 20.9)

Review of Basic Concepts

D.1 In the quadratic regression model $Y_i = \beta_0 + \beta_1 X + \beta_2 X^2 + \epsilon_i$, β_1 is called the _____ effect coefficient and β_2 is called the _____ effect coefficient.

D.2 The regression model in D.1 is a "linear" model because it is linear in the _____, even though it is not linear in the _____ _____.

D.3 Another curvilinear model is $Y_i = \beta_0 + \beta_1 X + \beta_2 X^2 + \beta_3 X^3 + \epsilon_i$, a cubic model. This model can be written in the form of the general linear regression model if we set $X_{i1} =$ _____, $X_{i2} =$ _____, and $X_{i3} =$ _____. ✓

»Problem 5

Refer to the situation in Problem 2. Based on the conclusion that there is no linear relation between number of visitors (Y) and age of zoo when number of exhibits (X_1) is already in the regression model (see Problem 2(c.), it was decided to drop the independent variable age of zoo from the model. It is now desired to determine whether or not number of exhibits also has a curvature effect. Computer output for this regression analysis, using the quadratic regression model (20.25) in your Text, is shown in Table 20.4, where $X_2 = X_1^2$.

Table 20.4

SELECTED COMPUTER OUTPUT FOR STUDY OF ANNUAL NUMBER
OF ZOO VISITORS USING QUADRATIC REGRESSION MODEL
(See Problem 6)

Variable	Reg. Coef.	Std. Dev.	t Stat
Constant	327.66700	70.9620	4.62
X_1	-5.53300	1.1900	-4.65
X_2	.04084	.0045	9.08

Analysis of Variance Table

Source	ss	df	MS
Regression	1730279	2	865139
Residual	69735	17	4102
Total	180001	19	

Variable	Specified Level
X_1	120
X_2	14400

Y Estimate = 251.80   Std. Dev. Mean Response = 19.19
                        Std. Dev. New Response = 66.86

a. i. State the appropriate alternatives for testing whether or not a curvature effect is present. _____.

    ii. Compute t*. _____ √

  iii. State the appropriate decision rule, using α = .01.

         _____

         _____

    iv. What conclusion should be reached? _____ Explain. _____

         _____ √

b. i. Using the computer output in Table 20.4, construct a 99
    percent confidence interval for the mean number of visitors
    among zoos with 120 exhibits. _____

   ii. Construct a 99 percent prediction interval for the number of
     visitors of a particular zoo with 120 exhibits. _____ √

»Problem 6

    A productivity analyst in a plant wished to examine the
relationship between the time (in hours) a worker takes to complete a
certain complex assembly (Y) and the number of previous such assemblies
completed by that worker (X). Test data for 20 workers appear in Table
20.5a. The analyst first fit these data to simple linear regression
model (19.1) in your Text, with results of the fitted model as shown in
Table 20.5b. Analysis of the residuals, however, seemed to suggest that
a model allowing for a curvilinear relationship might be more
appropriate. The analyst then transformed the independent variable with
a logarithmic transformation and fit the data to the following model: Y_i
= β_0 + $\beta_1 \log X_i$ + ϵ_i. Computer output for this analysis is shown in
Table 20.5c.

a. Based on the computer output in Table 20.5, does the model using
   $\log X_i$ appear to be more appropriate than that using X? _____
   Explain. _____ √

b. i. Using the computer output in Table 20.5c, construct a 90
    percent confidence for $E\{Y_h\}$ when X_h = 150 (i.e., when $\log X_h$
    = 2.17609). _____

   ii. Interpret this confidence interval. _____

     _____ _____ √

Table 20.5

REGRESSION RESULTS FOR STUDY OF ASSEMBLY TIMES
(See Problem 6)

(a)
Observations for Twenty Employees

Employee i	Previous Assemblies X_i	log X_i	Hours to Assemble Y_i
1	40	1.60206	7.7
2	65	1.81291	6.3
3	52	1.71600	7.0
4	141	2.14922	4.3
5	237	2.37475	3.9
6	93	1.96848	5.3
7	98	1.99123	5.0
8	35	1.54407	8.4
9	70	1.84510	6.1
10	157	2.19590	4.1
11	260	2.41497	3.8
12	123	2.08991	4.8
13	320	2.50515	3.7
14	103	2.01284	5.2
15	91	1.95904	5.6
16	43	1.63347	7.9
17	196	2.29226	4.1
18	308	2.48855	3.7
19	40	1.60206	7.3
20	52	1.71600	6.9

(b)
Selected Computer Output for
Regression Analysis Using X_1

Variable	Reg. Coef.	Std. Dev.	t Stat
Constant	7.402	.312	23.72
X_1	- .0146	.00202	-7.23

Analysis of Variance Table

Source	SS	df	MS
Regression	33.947	1	33.947
Residual	11.623	18	.646
Total	45.569	19	

R-sq = .745   R-sq(adj) = .731

(c)

Selected Computer Output for Regression Analysis Using log Xi

Variable	Reg. Coef.	Std. Dev.	T Stat.
Constant	15.165	.592	25.62
X_1	-4.815	.293	-16.43

Analysis of Variance Table

Source	SS	df	MS
Regression	42.717	1	42.717
Residual	2.852	18	.158
Total	45.569	19	

R-sq = .937   R-sq(adj) = .934

Variable	Specified Level
log X_1	2.17609

Y Estimate = 4.687   Std. Dev. Mean Response = .1034
Std. Dev. New Response = .4107

E. MODELING INTERACTION EFFECTS
(Text: Chapter 20, Section 20.10)

Review of Basic Concepts

E.1 Given the regression model $Y_i = \beta_0 + \beta_1 X_{i1} + \beta_2 X_{i2} + \beta_3 X_{i3} + \epsilon_i$, it is desired to incorporate an interaction effect between X_{i2} and X_{i3}. This is accomplished by expanding the model to include the interaction term _____ .

E.2 In the following model containing an interaction term, $Y = \beta_0 + \beta_1 X_{i1} + \beta_2 X_{i2} + \beta X_1 X_{i2} + \epsilon_i$, the effects of X_1 and X_2 are not additive. Thus, with this model, the effect of X_1 on _____ depends on the level of _____ .

E.3 Suppose a regression function is known to be $E(Y) = 20 + 15X_1 + 28X_2 - 2X_1X_2$. In this case independent variables X_1 and X_2 are said to _____ one another.

E.4 For the situation in E.3, if X_1 is fixed at the value $X_1 = 4$, the regression function relating Y to X_2 is _____ . Similarly, if X_1 is fixed at the value $X_1 = 8$, the regression function relating Y to X_2 is _____ .

E.5 Given the estimated response function $\hat{Y} = 150 - 25X_1 + 20X_2 + 10X_1X_2$, where X_2 is an indicator variable for gender ($X_2 = 1$ if male, $X_2 = 0$ otherwise), the estimated response function for males is _____ and for females is _____ . ✓

»Problem 7

Refer to the situation in Problem 4. The analyst here wished to expand the model discussed in Problem 4 to incorporate an interaction effect between indicator variable X_1 ($X_1 = 1$ if a snowbelt market, $X_1 = 0$ otherwise) and $X_2 =$ population density of a market. Using regression model 20.27) in your Text, the analyst obtained the computer output contained in Table 20.6, where $X_3 = X_1X_2$.

Table 20.6

SELECTED COMPUTER OUTPUT FOR STUDY OF FOUR
WHEEL DRIVE VEHICLES USING REGRESSION MODEL
CONTAINING INTERACTION TERM
(See Problem 7)

Variable	Reg. Coef.	Std. Dev.	t Stat
Constant	8.951	.466	19.21
X_1	7.580	.599	12.65
X_2	-.318	.067	-4.75
X_3	-.480 -	.089	-5.39

Analysis of Variance Table

Source	SS	df	MS
Regression	807.95	3	269.30
Residual	112.00	64	1.75
Total	919.95	67	

a. Conduct a test to determine whether or not an interaction effect
is present, using $\alpha = .01$. Specify the alternatives, determine
the value of the test statistic and the decision rule and state
the resulting decision. √

b. i. Plot the estimated response functions for snowbelt markets
($X_1 = 1$) and sunbelt markets ($X_1 = 0$) on one graph. (Note:
Population densities (in hundreds of persons per square mile)
in the data set fall roughly in the range $1 \leq X_2 \leq 12$.)

ii. Based on this graph, interpret the relationship between
percent of vehicles in a market that are four wheel drive and
the two independent variables. _____
                                                                    √

Answers to Chapter Reviews and Problems

A.1 a. $\beta_0 = 14.5$ is the intercept of the regression function. Equivalently, it is the mean response $E(Y)$ when $X_1 = 0$ and $X_2 = 0$.

b. $\beta_1 = 2.0$ is the change in $E(Y)$ when X_1 increases one unit while X_2 remains constant.

c. $\beta_2 = 12.5$ is the change in $E(Y)$ when X_2 increases one unit while X_1 remains constant.

A.2 plane

A.3 X_{i3} is the ith observation of the third independent variable in the model.

A.4 a. $p - 1 = 4$

b. $n - p = 45$

c. $n - 1 = 49$

A.5

Source	SS	df	MS
Regression	46488	2	23244.0
Error	43912	97	452.7
Total	90400	99	

A.6 $\hat{Y} = 642.25 - 14.60X_1 + 1.07X_2$

A.7 simple correlation

A.8 r_{Y3}

A.9 MSE; $43912/97 = 452.7$

A.10 0; 1; 1

A.11 $R^2 = 46488/90400 = .514$; no; R^2 can only remain the same or increase if a third independent variable is added to the model.

A.12 adjusted; R_a^2

Applied Statistics

A.13 1; SSE

A.14 $R_a^2 = 1 - (99/97)(43,912/90400) = 504$

A.15 $r^2_{Y3.12} = (1200 - 1050)/1200 = .125$

A.16 The error sum of squares is reduced 12.5 percent by adding X_3 to the regression model when X_1 and X_2 are already in the model.

Problem 1

a. $E\{Y\} = \beta_0 + \beta_1 X_1 + \beta_2 X_2$

b. i. $\hat{Y} = -74.189 + 10.626X_1 + 25.973X_2$

   ii. $b_2 = 25.973$ is a point estimate of the increase in the mean maximum pounds of pressure per square centimeter for each unit increase in density of material while thickness of material is held constant.

c. $R^2 = 5958.4/6446.0 = .924$. Thus, the variability in toy strength (in maximum pounds of pressure per square centimeter) is reduced 92.4 percent when thickness of material and density of material are considered in the regression model.

d. $r^2Y_{2.1} = (1405.8 - 487.6)/1405.8 = .6 - 53$. Thus, the error sum of squares is reduced 65.3 percent by adding density of material (X_2) to the regression model when thickness of material (X_1) is already in the model.

Problem 2

a.   Alternatives: $H_0: \beta_1 = \beta_2 = 0$; H_1: not both $\beta_1 = 0$ and $\beta_2 = 0$

   F*:   MSR = 698337; MSE = 23726;
   Thus, F* = 698337/23726 = 29.433.

   Decision Rule: $F(.99; 2, 17) = 6.11$. Thus the decision rule is:
   If $|F*| \leq 2.898$, conclude H_0;
   If $|F*| > 2.898$, conclude H_1.

Decision: Since F* = 29.433 > 6.11, conclude H_1, there is a linear regression relation between number of visitors and the independent variables, number of exhibits (X_1) and age of zoo (X_2).

b.   Alternatives: H_0: $\beta_1 = 0$; H_1: $\beta_1 \neq 0$

     t*: $b_1 = 5008$; s(b1) = .692. Thus, t* = 7.24.

Decision Rule: t(.995; 17) = 2.898. Thus, the decision rule is:
If $|t*| \leq 2.898$, conclude H_0;
If $|t*| > 2.898$, conclude H_1;

Decision: Since $|t*| - 7.24 > 2.898$, conclude $H_1(\beta_1 \neq 0)$, that, with age of zoo (X_2) already in the model, there is a linear relation between number of visitors (Y) and number of exhibits (X_1).

c.   Alternatives: H_0: $\beta_2 = 0$; H_1: $\beta_2 \neq 0$

     t*:   t* = -.32

Decision Rule: If $|t*| \leq 2.898$, conclude H_0;
If $|t*| > 2.898$, conclude H_1;

Decision: Since $|t*| = .32 \leq 2.898$, conclude $H_0(\beta_2 = 0)$, that, with number of exhibits (X_1) already in the model, there is no linear relation between number of visitors (Y) and age of zoo (X_2). (Note: This conclusion does not necessarily mean that Y and X_2 would be unrelated linearly if X_2 were to be considered alone in the model. It only means that when X_1 is in the model, the marginal contribution of X_2 in further reducing the error sum of squares is insignificant.)

Problem 3

a. i. $b_1 = 10.626$; $s\{b_1\} = 1.872$; $t(.975; 7) = 2.365$. The confidence limits are $10.626 \pm 2.365(1.872)$. Thus, the 95 percent confidence interval for β_1 is $6.19g \leq \beta_1 \leq 15.053$.

   ii. $b_2 = 25.973$; $s\{b_2\} = 7.154$; $t(.975; 7) = 2.365$. Thus, the 95 percent, confidence interval for β_2 is $9.054 \leq \beta_2 \leq 42.892$.

  iii. With 95 percent confidence, we can state that a 1 mil increase in thickness of material (X_1) is associated with an increase between 6.199 and 15.053 in mean maximum pounds of pressure per square centimeter, while density of material (X_2) is held fixed. Similarly, with 95 percent confidence, we can state that an increase of 1 in the index of density of material (X_2) is associated with an increase between 9.054 and 42.892 in mean maximum pounds of pressure per square centimeter, while thickness of material (X_1) is held fixed.

b. i. $\hat{Y}_h = -74.189 + 10.626(4) + 25.973(3.4) = 56.623$; $s\{\hat{Y}_h\}$; 3.23; $t(.99, 7) = 2.998$. The confidence limits are $56.623 \pm 2.998(3.236)$. Thus, the 98 percent confidence interval for $E\{Y_h\}$ when $X_{h1} = 4$ and $X_{h2} = 3.4$ is $46.921 \leq E\{Y_h\} \leq 66.325$.

   ii. $\hat{Y} = 56.623$; $s\{Y_{h(new)}\} = 8.951$; $t(.99; 7) = 2.998$. The prediction limits are $56.623 \pm 2.998(8.951)$. Thus, the 98 percent prediction interval for $Y_{h(new)}$ when $X_{h1} = 4$ and $X_{h2} = 3.4$ is $29.788 \leq Y_{h(new)} \leq 83.458$.

  iii. Yes. As in simple linear regression, for a specified level of the independent variable(s), the prediction interval for a new response with confidence coefficient $1 - \alpha$ is necessarily wider than the corresponding confidence interval for a mean response. This is so, since, as in simple linear regression, $s^2\{Y_{h(new)}\} = MSE + s^2\{\hat{Y}_h\} > s^2\{\hat{Y}_h\}$.

C.1 full-time; part-time

C.2 dummy or binary

C.3 slope; intercepts

C.4 $E\{Y\} = 35 + 6X_1$; $E\{Y\} = 50 + 6X_1$

C.5 larger

C.6 three

C.7 There are several ways to define the set of three indicator variables. If we denote them, say X_1, X_2 and X_3 and let the class "private home" be the reference class, one appropriate set is:

$X_1 \equiv 1$ if small apartment building, 0 otherwise

$X_2 \equiv 1$ if large apartment building, 0 otherwise

$X_3 \equiv 1$ if condominium, 0 otherwise

Problem 4

a.   Alternatives: H_0: $\beta_1 = \beta_2 = 0$; H_1: not both $\beta_1 = 0$ and $\beta_2 = 0$

            F*:   MSR/MSE = 378.6/2.5 = 151.4.

       Decision Rule: $F(.99; 2, 65) = 4.96$. (Note: This percentile was obtained from Table C-4 via linear interpolation.)
If $|F*| \le 4.96$, conclude H_0;
If $|F*| > 4.96$, conclude H_1.

          Decision: Since F* = 151.4 > 4.96, conclude H_1, there is a relationship between percent of vehicles that are four wheel drive (Y) and the independent variables, snowbelt versus sunbelt market (X_1) and population density (X_2).

b.   Alternatives: H_0: $\beta_1 = 0$; H_1: $\beta_1 \neq 0$

            t*: $t* = b_1/s\{b_1\} = 4.878/.391 = 12.48$

       Decision Rule: $t(.995; 65) = 2.656$. The decision rule is:
$|t*| \le 2.656$, conclude H_0;
$|t*| > 2.656$, conclude H_1.

Decision: Since $|t^*| = 12.48 > 2.656$, conclude $H_1(\beta_1 \neq 0)$, that, with population density (X_2) already in the model, there is a relation between percent of vehicles that are four wheel drive (Y) and snowbelt versus sunbelt market (X_1).

c. i. $\hat{Y} = 10.538 + 4.878X_1 - .588X_2$

ii. $b_1 = 4.878$; thus, for any given population density, the mean percent of vehicles in a snowbelt market that are four wheel drive is estimated to be 4.878 percent points higher than that in a sunbelt market.

iii. $b_2 = -.588$; Thus, the mean percent of vehicles in a market that are four wheel drive is estimated to decrease .588 percent points for each unit increase in population density (i.e., for each increase of 100 persons per square mile).

D.1 linear; curvature

D.2 parameters, independent variable

D.3 $X_{i1} = X_i$; $X_{i2} = X_i{}^2$; $X_{i3} = X_i{}^3$

Problem 5

a. i. $H_0: \beta_2 = 0$; $H_1: \beta_2 \neq 0$

ii. $t^* = .04084/.0045 = 9.08$

iii. $t(.995; 17) = 2.898$. Thus, the decision rule is:
$|t^*| \leq 2.898$, conclude H_0;
$|t^*| > 2.898$, conclude H_1.

iv. Since $|t^*| = 9.08 > 2.898$, conclude $H_1(\beta_2 \neq 0)$, there is a curvature effect in the regression relation between number of visitors and the independent variable, number of exhibits.

b. i. $X_{h1} = 120$; $X_{h2} = (120)^2 = 14400$; $\hat{Y}_h = 327.667 - 5.533(120) + .04084(14400) = 251.80$; $s\{\hat{Y}_h\} = 19.19$; $t(.995; 17) = 2.898$. The confidence limits are $251.80 \pm 2.898(19.19)$. Thus, the 99 percent confidence interval for $E\{Y_h\}$) when $X_{h1} = 120$ and $X_{h2} = 14400$ is $196.19 < E\{Y_h\} \leq 307.41$.

iii. $\hat{Y}_h = 251.80$; $s\{Y_{h(new)}\} = 66.86$; $t(.995; 17) = 2.898$. Thus, the 99 percent prediction interval for $Y_{h(new)}$ when $X_{h1} = 120$ and $X_{h2} = 14400$ is $58.04 \leq Y_{h(new)} \leq 445.56$.

Problem 6

a. Yes; residual variability in assembly time is considerably less when $logX_i$ is used (i.e., SSE = 2.852, MSE = .158) than when X_i is used (i.e., SSE = 11.623; MSE = .646, indicating the sample cases fit the model using $logX_i$ better than that using X_i. This improved fit can also be seen in the respective value of R_a^2 which increased from $R_a^2 = .731$ in the model using X_i to $R_a^2 = .934$ in the model using $logX_i$.

b. i. $\hat{Y}_h = 15.165 - 4.815(2.17609) = 4.687$; $s\{\hat{Y}_h\} = .1034$; $t(.95; 18) = 1.734$. The confidence limits are $4.687 \pm 1.734(.1034)$. Thus, the 90 percent confidence interval for $E\{Yh\}$ when $X_h = 150$ (i.e., when $logX_h = 2.17609$) is $4.508 \leq E\{Y_h\} \leq 4.866$.

ii. With 90 percent confidence, the mean time to complete this assembly among workers who have previously completed 150 such assemblies is between 4.508 hours and 4.866 hours.

E.1 $\beta_4 X_{i2} X_i 3$

E.2 Y; X_2

E.3 interfere with

E.4 $E\{Y\} = 20 + 15(4) + 28X_2 - 2(4)X_2 = 80 + 20X_2$; $E\{Y\} = 140 + 12X_2$

E.5 $\hat{Y} = 250 - 25X_1 + 20(1) + 10X_1(1) = 170 - 15X_1$; $\hat{Y} = 150 - 25X_1$

»Problem 7

a. Alternatives: H_0: $\beta_3 = 0$; H_1: $\beta_3 \neq 0$

    t*: $t* = b_1/s\{b_1\} = 4.80/.089 = -5.39$

    Decision Rule: $t(.995; 64) = 2.657$. Thus the decision rule is:
    $|t*| \leq 2.657$, conclude H_0;
    $|t*| > 2.657$, conclude H_1.

Decision: Since $|t*| = 5.39 > 2.657$, conclude $H_1(\beta_3 \neq 0)$, that there is an interaction effect between indicator variable X_1 ($X_1 = 1$ if a snowbelt market, $X_1 = 0$ otherwise) and $X_2 =$ population density of a market.

b. i.

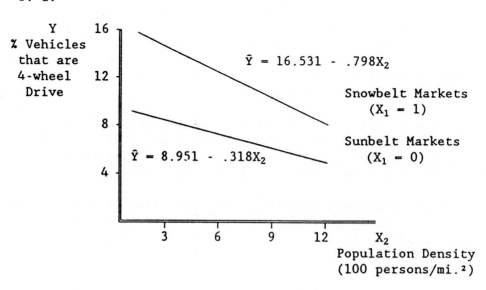

Figure 20.1

ii. As shown in Figure 20.1, in snowbelt markets the estimated mean percent of vehicles that are four wheel drive declines from about 16.5 percent for markets with one hundred persons per square mile to just 7.0 percent for markets with 12 hundred persons per square mile. In sunbelt markets the percent declines from just under 9.0 percent to just over 5.0 percent. Also, for any given population density within the range of population densities studied, the estimated mean percent of vehicles that are four wheel drive is greater in snowbelt than in sunbelt markets, but that disparity decreases as population density increases.

21

ANALYSIS OF VARIANCE

A. ANALYSIS OF VARIANCE MODEL
(Text: Chapter 21, Sections 21.1 and 21.2)

Review of Basic Concepts

A.1 An experiment was designed to test the effect of type of package on the shelf life (in weeks) of a particular snack food. Three different packages were examined; a heat-sealed waxed paper, a plastic tub with screw-on lid, and a vacuum sealed tin. Identify the dependent variable and independent variable in this study.

a. Dependent Variable _____

b. Independent Variable _____

A.2 In ANOVA terminology, the independent variable is called a(n) _____.

A.3 For the situation in A.1, the three different package designs tested are called _____ _____ or _____.

A.4 A study was designed to examine the effect of price of a product on sales volume. The following four price levels were utilized in the study; $3.49, $3.79, $3.99, and $4.39. Each of the four price levels was used in a sample of two test cities and sales volume was measured in terms of number of units sold per 1,000 population. The ANOVA model assumes that sales volume in each city receiving the first treatment (i.e., a price of $3.49) is composed of two parts, a constant term which is denoted by _____ and an error term which is denoted by _____.

A.5 For the situation in A.4, the error terms are assumed to be independent $N(0, \sigma^2)$. Therefore, observed sales volumes under the first treatment, denoted by _____ constitute random variables which are _____ of one another and which follow a probability distribution that is _____ with a mean of _____ and variance _____.

A.6 An ANOVA model was used to study the effect of household income on household spending for recreation. In this study, three income classes (lower, middle, upper) were employed. Is this a single-factor or multifactor ANOVA? _____ Explain.

_____

A.7 For the situation in A.6, is household income an experimental or observational factor? _____ Explain. _____

_____

A.8 In the ANOVA model, the probability distributions for the treatments are all assumed to be normal with constant variance. Therefore, the only way in which treatments may differ from one another is with respect to their _____ ✓

B. TEST FOR EQUALITY OF TREATMENT MEANS
(Text: Chapter 21 , Section 21.3)

Review of Basic Concepts

B.1 Identify the symbol used to denote each of the following in single-factor analysis of variance.

a. The number of sample observations for treatment 2. _____

b. The total number of sample observations in the study. _____

c. The sum of the sample observations for treatment 2. _____

d. The mean of the sample observations for treatment 2. _____

e. The overall mean for all the sample observations. _____

B.2 The following data resulted from an ANOVA study with three treatments.

Observation	Treatment (j)		
i	1	2	3
1	24	33	25
2	28	36	31
3	26	36	22

Determine each of the following for this study.

a. $n_1 =$ _____ ; $n_2 =$ _____ ; $n_3 =$ _____ ; $n_T =$ _____ .

b. $\overline{Y}_1 =$ _____ ; $\overline{Y}_2 =$ _____ ; $\overline{Y}_3 =$ _____ ; $\overset{=}{Y} =$ _____ .

B.3 The total sum of squares, denoted by SS ___ , is based on the deviations of the observations Y_{ij} around the _____ _____ , while the error sum of squares, denoted by SS ___ , is based on the deviations of the observations Y_{ij} around the _____ _____ .

B.4 The difference SSTO - SSE is called the _____ sum of squares and is denoted by SS ___ . This sum of squares is based on the deviations of the _____ _____ around the _____ _____ .

B.5 Determine the degrees of freedom associated with each sum of squares in a study involving three treatments, each with six observations.

a. SSTO: _____

b. SSTR: _____

c. SSE: _____

B.6 For the situation in B.5, SSTO = 3400 and SSE = 2400. Therefore, SSTR = _____ . Further, MSTR = _____ and MSE = _____ .

B.7 In analysis of variance, MSTR is called the _____ _____ _____ and MSE is called the _____ _____ _____ .

B.8 The F test for equality of treatment means is based on the test statistic F* = _____/_____ . For the situation in B.5 and B.6, F* = _____ .

B.9 Regarding the relative magnitude of E{MSTR} versus E{MSE}, when the treatment means μ_j are all equal, E{MSTR} _____ E{MSE}. When the treatment means μ_j differ, E{MSTR} _____ E{MSE}. Thus, relatively _____ values of F* lead to conclusion H1 (not all μ_j's are equal).

B.10 Consider the following ANOVA table, which resulted from a study involving five treatments with three observations for each treatment.

Source	SS	df	MS
Treatments	400	4	100
Error	750	10	75
Total	1150	14	

If it is desired to control the α risk at .01, the action limit of the decision rule is based on the 99th percentile of the F distribution with _____ and _____ degrees of freedom. Thus, we obtain F(.99, _____, _____) = _____.

B.11 For the situation in B.10, state the appropriate decision rule.

B.12 For the situation in B.10 and B.11, what conclusion should be reached? _____ Why? _____ √

»Problem 1

A tire manufacturer desired to compare the durability of three types of construction material for tires; experimental material A, experimental material B, and conventional material C. A test was designed in which three tires of each type of material were subjected to wear tests in a laboratory. The results of this test, in thousands of miles of wear, were as follows:

	Treatment (j)		
Observation	1	2	3
i	(MaterialA)	(MaterialB)	(MaterialC)
1	24	39	25
2	30	38	24
3	24	43	32

a. Given the results in the preceding table, determine each of the following.

   i. $\bar{Y}_1$ = _____

   ii. $\bar{Y}_2$ = _____

   iii. $\bar{Y}_3$ = _____

   iv. $\bar{\bar{Y}}$ = _____ √

b. Obtain the sums of squares for this study using (21.7), (21.9), and (21.10) in your Text.

   i. SSTO = _____

   ii. SSE = _____

   iii. SSTR = _____ √

c. Complete the following ANOVA table for this study. √

Source	SS	df	MS
Treatments	____	____	____
Error	____	____	____
Total	____	____	____

d. i. State the appropriate alternatives to test whether or not mean wear is the same for all three materials. _____

   ii. Determine the test statistic for this test. _____ √

    iii. Using $\alpha = .05$, obtain the appropriate decision rule for this
         test. _____

    iv. What conclusion should be reached? _____ Explain. _____
      _____ √

  e. Does your conclusion in (d.iv.) necessarily imply that the means
    for the two experimental materials are not equal? _____
    Explain. _____ √

C. ANALYSIS OF TREATMENT EFFECTS
(Text: Chapter 21, Section 21.4)

Review of Basic Concepts

C.1 If interest centers on a particular treatment mean μ_j, we estimate
this mean with a confidence interval based on the t distribution
with _____ degrees of freedom.

C.2 In a particular ANOVA study, $MSE = 75.0$ and $n_2 = 3$. Thus, the
estimated variance of $\bar{Y}_2$ is $s^2(\bar{Y}_2) =$ _____ and the estimated
standard deviation of $\bar{Y}_2$ is $s(\bar{Y}_2) =$ _____.

C.3 In an ANOVA study with three treatments, $\bar{Y}_1 - 87.0$, $\bar{Y}_2 - 61.5$, and
$\bar{Y}_3 - 73.5$. It is desired to estimate $\mu_3 - \mu_2$. An unbiased
estimator of $\mu_3 - \mu_2$ is _____ which in this case, equals
_____.

C.4 For the situation in C.3, $MSE = 10.0$ and there are four
observations for each treatment. Therefore, $s^2(\bar{Y}_3 - \bar{Y}_2) =$
_____ and $s(\bar{Y}_3 - \bar{Y}_2) =$ _____. Further, if this is the
only pairwise comparison to be made in this study and a 90 percent
confidence interval is desired, the confidence interval is
constructed using $t(1-\alpha/2; n_T-r) =$ _____.

C.5 For the situation in C.3 and C.4, the confidence limits are
_____ ± _____. Thus, the 90 percent confidence interval
for $\mu_3 - \mu_2$ is _____ $\leq \mu_3-\mu_2 \leq$ _____.

C.6 For an analysis of variance study, the 95 percent confidence
interval for $\mu_4 - \mu_1$ is $-\$12 \leq \mu_4-\mu_1 \leq -\$4$. Interpret this
confidence interval. _____
_____

C.7 Three pairwise comparisons are to be made in a study involving three treatments, each with a 99% confidence interval. The probability that all three confidence intervals will be simultaneously correct is at least _____ and, hence, their joint confidence coefficient is at least _____ percent.

C.8 In an ANOVA study with five treatments, four simultaneous pairwise comparisons are to. be made. If a joint confidence coefficient of at least 96 percent is desired, the confidence interval for each of the four pairwise comparisons is constructed with a(n) _____ percent confidence coefficient.

C.9 For the situation in C.8, if there are five observations for each treatment, each confidence interval is constructed using the _____ percentile of the t distribution with _____ degrees of freedom. √

   »**Problem 2**

   Refer to the situation in Problem 1. Recall that the three treatments refer to experimental material A, experimental material B, and conventional material C, respectively, and that $n_1 = n_2 = n_3 = 3$. The sample treatment means are $Y_1 = 26.0$, $Y_2 = 40.0$, and $Y_3 = 27.0$ thousand miles of wear. Finally, MSTR = 183.00 and MSE = 12.67.·

   a. Construct a 95 percent confidence interval for mean wear with conventional material C. _____ √

   b. The analyst in this study wishes to compare mean wear of experimental material B with mean wear of experimental material A and also to compare mean wear of conventional material C with mean wear of each experimental material, using a joint confidence coefficient of 94 percent for the three simultaneous pairwise comparisons.

      i. Construct the appropriate confidence interval for $\mu_2 - \mu_1$.
         _____ √

      ii. Construct the appropriate confidence interval for $\mu_3 - \mu_1$.
          _____ √

      iii. Construct the appropriate confidence interval for $\mu_3 - \mu_2$.
           _____ √

c. Interpret your findings in (b.). _____

_____

_____

D. RESIDUAL ANALYSIS
(Text: Chapter 21, Sections 21.5 and 21.6)

Review of Basic Concepts

D.1 The residual for the ith observation of the jth treatment, denoted by ____, is defined aa the difference between the observed response for the ith observation in the jth treatment, denoted by _____, and the sample mean for the jth treatment, denoted by _____.

D.2 The $n_1 = 4$ observations for the first treatment in an ANOVA study were $Y_{11} = 50$, $Y_{21} = 56$, $Y_{31} = 51$, and $Y_{41} = 59$. Therefore, $Y_1 =$ _____.

D.3 For the situation in D.2, the residuals for treatment 1 are $e_{11} =$ _____, $e_{21} =$ _____, $e_{31} =$ _____, $e_{41} =$ _____. Thus, $\sum\limits_{i} e_{i1} =$ _____.

D.4 The following residual plots resulted from an ANOVA study involving three treatments with twelve observations for each treatment.

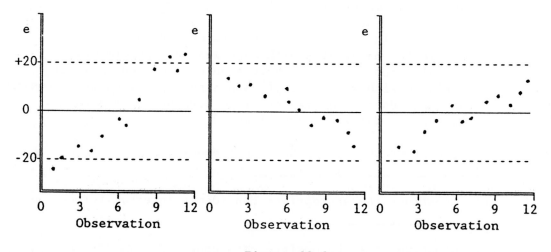

Figure 21.1

Do the residuals in these plots provide evidence of a departure from the assumptions of ANOVA model (21.1) in your Text? _____ Explain. _____

D.5 For an ANOVA study involving four treatments, each with eight observations, inferences based on the F test for equality of treatment means are seriously affected if the error terms are not _____ but are seriously affected if the error terms have unequal _____.

D.6 Examine the following residual plots, which resulted from an ANOVA study with two treatments and 20 observations for each treatment.

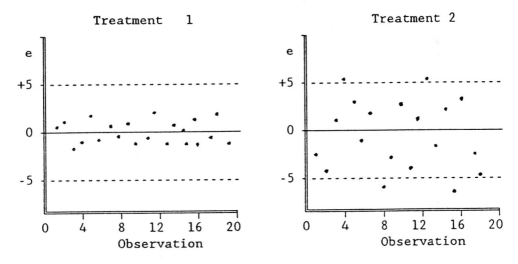

Figure 21.2

Do the error terms within each treatment appear to be independent? _____ Explain. _____ _____

D.7 For the situation in D.6, do the error variances for the two treatments appear to be equal? _____ Explain. _____

D.8 An experiment was conducted using four treatments with eight observations for each treatment. The following residual plot is a combined residual plot of all $n_T = 32$ observations, plotted in the order in which the observations were obtained.

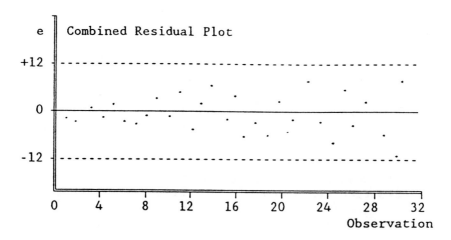

Figure   21.3

Which assumption regarding the error terms in model (21.1) in your Text does not appear to be satisfied in this situation? _____ Explain. _____

_____

_____ √

Answers to Chapter Reviews and Problems

A.1 a. Dependent variable is shelf life (in weeks).
      b. Independent variable is type of package.

A.2 factor

A.3 factor levels; treatments

A.4 μ_1; ϵ_{i1}

A.5 Y_{i1}; independent; normal; μ_1; σ^2

A.6 Single-factor. In this study there is just one independent variable or factor, household income, which has three levels (lower, middle, and upper).

A.7 Observational factor. Income class is a characteristic of a household which the analyst can observe or measure. However, income class cannot be randomly assigned to households.

A.8 means (μ_j)

B.1 a. n_2

    b. n_T

    c. $\displaystyle\sum_{i=1}^{n_2} Y_{i2}$

    d. $\overline{Y}_2$

    e. $\overline{\overline{Y}}$

B.2 a. $n_1 = 3$; $n_2 = 3$; $n_3 = 3$; $n_T = 9$

    b. $\overline{Y}_1 = 26.0$; $\overline{Y}_2 = 35.0$; $\overline{Y}_3 = 26.0$; $\overline{\overline{Y}} = 29.0$

B.3 SSTO; overall mean $(\overline{\overline{Y}})$; SSE; treatment means $(\overline{Y}_j)$

B.4 treatment, SSTR; treatment means $(\overline{Y}_j)$; overall mean $(\overline{\overline{Y}})$

B.5 a. $18 - 1 = 17$

    b. $3 - 1 = 2$

    c. $18 - 3 = 15$

B.6 SSTR $= 3400 - 2400 = 1000$; MSTR $= 1000/2 = 500$; MSE $= 2400/15 = 160$

B.7 treatment mean square; error mean square

B.8 MSTR/MSE; $F^* = 500/160 = 3.125$

B.9 E(MSTR) $=$ E(MSE); E(MSTR) $>$ E(MSE); large

B.10 4; 10; $F(.99; 4, 10) = 5.99$

B.11 If $F^* \leq 5.99$, conclude H_0; if $F^* > 5.99$, conclude H_1

B.12 $F^* = 100/75 = 1.33$. Since $F^* = 1.33 < 5.99$, conclude $H_0(\mu_1 = \mu_2 = \mu_3 = \mu_4 = \mu_5)$.

Problem 1

a. i. $\bar{Y}_1 = (24 + 30 + 24)/3 = 26.0$

 ii. $\bar{Y}_2 = (39 + 38 + 43)/3 = 40.0$

 iii. $\bar{Y}_3 = (25 + 24 + 32)/3 = 27.0$

 iv. $\bar{\bar{Y}} = (24+30+\cdots+24+32)/9 = 31.0$

b. i. $SSTO = (24\text{-}31)^2 + (30\text{-}31)^2 + (24\text{-}31)^2 + (39\text{-}31)^2 +$
    $(38\text{-}31)^2 + (43\text{-}31)^2 + (25\text{-}31)^2 + (24\text{-}31)^2 + (32\text{-}31)^2 = 442$

 ii. $SSE = (24\text{-}26)^2 + (30\text{-}26)^2 + (24\text{-}26)^2 + (39\text{-}40)^2 + (38\text{-}40)^2 +$
    $(43\text{-}40)^2 + (25\text{-}27)^2 + (24\text{-}27)^2 + (32\text{-}27)^2 = 76$

 iii. $SSTR = 442 - 76 = 366$

c.

Source	SS	df	MS
Treatments	366.0	2	183.00
Error	76.0	6	12.67
Total	442.0	8	

d. i. $H_0: \mu_1 = \mu_2 = \mu_3$; H_1: not all μ_j's are equal

 ii. $F* = 183.00/12.67 = 14.44$

 iii. $F(.95; 2, 6) = 5.14$. Thus, the decision rule is:
    If $F* \leq 5.14$, conclude H_0;
    If $F* > 5.14$, conclude H_1.

 iv. Since $F* = 14.44 > 5.14$, conclude H_1 (not all μ_j's are
    equal), mean wear is not the same for all three materials.

e. No; the conclusion is that not all of μ_1, μ_2, and μ_3 are equal.
   It could, however, be the case that $\mu_1 = \mu_2 \neq \mu_3$.

C.1 $n_T - r$

C.2 $s^2\{\bar{Y}_2\} = 75/3 = 25.0$; $s\{\bar{Y}_2\} = 5.0$

C.3 $\bar{Y}_3 - \bar{Y}_2 = 73.5 - 61.5 = 12.0$

C.4 $s^2(\overline{Y}_3 - \overline{Y}_2) = (10/4) + (10/4) = 5.0$; $s(\overline{Y}_3 - \overline{Y}_2) = 2.236$; $t(.95; 9)$
= 1.833

C.5 $12.0 \pm 1.833(2.236)$; $7.901 \le \mu_3 - \mu_2 \le 16.099$

C.6 With 95 percent confidence, we can state that the mean response for treatment 4 is between 4.0 and 12.0 dollars less than the mean response for treatment 1.

C.7 .97; 97

C.8 99

C.9 $1 - \alpha/2m = .995$ or 99.5th percentile; $25 - 5 = 20$

Problem 2

a. $\overline{Y}_3 = 27.0$; $s^2(\overline{Y}_3) = 12.67/3 = 4.223$; $s(\overline{Y}_3) = 2.055$; $t(.975; 6) = 2.447$. The confidence limits are $27.0 \pm 2.447(2.055)$. The 95 percent confidence interval for μ_3 is $21.971 \le \mu_3 \le 32.029$.

b. i. $\overline{Y}_2 - \overline{Y}_1 = 40.0 - 26.0 = 14.0$; $s^2(\overline{Y}_2 - \overline{Y}_1) = 12.67/3 + 12.67/3 = 8.447$; $s(\overline{Y}_2 - \overline{Y}_1) = 2.906$; $1 - \alpha = .94$; $1 - \alpha/2m = .99$; $t(.99; 6) = 3.143$. The confidence limits are $14.0 \pm 3.143(2.906)$. The appropriate confidence interval for $\mu_2 - \mu_1$ is $4.866 \le \mu_2 - \mu_1 \le 23.134$.

   ii. $\overline{Y}_3 - \overline{Y}_1 = 27.0 - 26.0 = 1.0$; $s(\overline{Y}_3 - \overline{Y}_1) = 2.906$; $t(.99; 6) = 3.143$. Thus, the appropriate confidence interval for $\mu_3 - \mu_1$ is $-8.134 \le \mu_3 - \mu_1 \le 10.134$.

   iii. $\overline{Y}_3 - \overline{Y}_2 = 27.0 - 40.0 = -13.0$; $s(\overline{Y}_3 - \overline{Y}_2) = 2.906$; $t(.99; 6) = 3.143$. Thus, the appropriate confidence interval for $\mu_3 - \mu_2$ is $-22.134 \le \mu_3 - \mu_2 \le -3.866$.

c. Experimental material B (treatment 2) appears to be more durable than either experimental material A (treatment 1) or conventional material C (treatment 3), but it is not clear if experimental material A is more or less durable than conventional material C. The overall of joint confidence we have that all three comparisons are simultaneously correct is 94 percent.

D.1 e_{ij}, Y_{ij}, $\overline{Y}_j$

D.2 $\overline{Y}_1 = 54$

D.3 $e_{11} = -4$; $e_{21} = 2$; $e_{31} = -3$; $e_{41} = 5$; $\sum\limits_{i} e_{i1} = (-4) + 2 + (-3) + 5 = 0$

D.4 Yes. If the error terms were independent, the residuals for each treatment would tend to be scattered at random around 0. In this study, however, the residuals for treatments 1 and 3 show a marked upward pattern while those for treatment 2 show a marked downward pattern.

D.5 independent; variances

D.6 Yes; within each treatment the residuals tend to be scattered at random around 0.

D.7 No; the residual plots indicate that the variance for treatment 1 is less than the variance for treatment 2.

D.8 Constancy of error variance; the residuals tend to become progressively more variable.

22

QUALITY CONTROL

A. CONTROL CHARTS
(Text: Chapter 22, Section 22.3)

Review of Basic Concepts

A.1 In a quality control chart, UCL denotes the _____
_____ _____ and the LCL denotes the
_____ _____ _____.

A.2 In a control chart for a process mean, the horizontal axis is used
to record _____ and the vertical axis to record the
values of the _____ _____.

A.3 Control charts are usually constructed with UCL and LCL set at
_____ standard deviations above and below the process mean.
Thus, they are based on a very small _____ risk. In the
context of control charts, what does this risk measure? _____
_____

A.4 A particular process is set to produce ball bearings with a mean
diameter of 1.5 cm and a standard deviation of .05 cm. Random
samples of n = 100 ball bearings are selected hourly to test
whether or not the process mean is in control. The control chart
for this situation is constructed with UCL = 1.515 and LCL = 1.485.
The sample mean for the last sample was $\bar{X}$ = 1.479. Is the
process is in control or out of control? _____ Explain.
_____

A.5 What managerial action is implied by the conclusion that a process
is out of control? _____
_____
_____

A.6 Of what managerial importance is it to identify an upward or downward trend in a process mean when the process is still within the control limits? _____

_____

A.7 A process that manufactures incandescent lamps is subject to quality control testing using control charts. When under control, three percent of the lamps manufactured by the process are defective. Hourly random samples of n = 400 lamps are selected to determine whether or not the process is under control. For this situation, LCL = _____ and UCL = _____.

A.8 The number of orders per day that do not receive "same-day" attention an order filling process is known to follow a Poisson distribution. Based on past experience, the mean number of orders that do not receive such attention is λ_0 = 8.0 when the process is under control. A control chart is to be used to monitor this process. For this situation, LCL = _____ and UCL = _____.

A.9 For the situation in A.4, the process is reset at 8:00 a.m. Monday. The follow results were recorded for the next four samples:

i	Time	$\overline{X}$	C_k
1	9:00 a.m.	1.503	_____
2	10:00 a.m.	1.507	_____
3	11:00 a.m.	1.492	_____
4	12:00 p.m.	1.504	_____

It is desired to plot these results on a cusum chart beginning with the 9:00 a.m. sample. Compute the relevant cumulative sums, C_k that would be plotted on such a chart.

A.10 For the situation in A.4 and A.9, the resulting cusum chart through Wednesday showed just mild fluctuations about 0 throughout the three day period. What conclusion should be reached about the operation of the process during this period? _____

_____ √

»Problem 1

A manufacturer of pre-packaged cookies uses a machine which measures and dispenses dough for a particular flavor of cookies. When the machine operates properly, dough is dispensed in such a quantity that the weights of baked cookies are approximately normal with μ = .350 ounces and σ = .015 ounces. The production manager wishes to construct a control chart to use in monitoring the mean weight of cookies over time. Plans are made to select random samples of n = 25 cookies at 10:00 a.m., 12:00 p.m., 2:00 p.m., and 4:00 p.m. each working day. On the basis of the mean weight of each sample, a decision will be made as to whether or not the process mean is in control.

a. State the alternatives to be used for each test of whether or not the process mean is in control. _____

b. Determine the control limits to be used in constructing the control chart. LCL = _____; UCL = _____ √

c. Construct the appropriate control chart for this situation for a five-day period. √

d. Suppose the production manager adjusts the dispensing machine on Monday, so that it functions properly and begins using the control chart. Sample results taken over the next three days, listed in order from 10:00 a.m. Monday to 4:00 p.m. Wednesday are: .351, .348, .353, .352, .353, .354, .351, .353, .355, .356, .358, and .361.

   i. Plot these sample means on the same control chart in part (c.).

   ii. Was the process in control during this period? _____ Explain. _____

   _____ √

e. The machine is reset on Thursday morning. Samples for the next two days yield sample means of .348, .352, .350, .349, .351, .352, .355, and .356.

   i. Plot these sample means on the same control chart.

ii. Was the process in control during these two days? _____
Explain. _____
_____ √

f. i. Construct a cusum chart based on the sample results from
10:00 a.m. Monday through 4:00 p.m. Friday. (Note: Be sure
your cusum chart reflects that the machine was reset on
Thursday morning; i.e., begin anew cumulating departures
commencing with the 10:00 a.m. Thursday sample.)

ii. Does your cusum chart provide information about this process
that would be useful to the production manager? _____
Explain. _____
_____ √

B. ACCEPTANCE SAMPLING
(Text: Chapter 22 , Section 22.4)

Review of Basic Concepts

B.1 A particular lathe is set to produce fabricating parts for
subsequent assembly at the rate of 1,500 parts per hour. If the
proportion of defective parts in any hour's production is greater
than .03, the entire production for that hour should be rejected
and sent to a separate department for 100 percent inspection.
Otherwise, the lot of 1,500 parts should be crated and sent to
inventory. The company selects a simple random sample of n = 50
parts from each lot and employs the following decision rule: If 3
or more parts are defective the lot is rejected. Otherwise, the
lot is sent to inventory. In this situation, the acceptable
quality level is AQL = _____.

B.2 For the situation in B.1, the acceptance number is _____ and
the rejection number is _____.

B.3 For the situation in B.1, the α risk is the probability that a lot
of _____ quality is _____. The α risk is often
referred to as the _____ risk.

B.4 For the situation in B.1, the ß risk is the probability that a lot
of _____ quality is _____. The ß risk is often
referred to as the _____ risk.

B.5 In an acceptance sampling situation, interest centers on the probability of accepting a lot with proportion p = .10 defectives. This probability is denoted by _____. A plot of this probability along with similarly defined probabilities for different values of p is called the _____ _____ _____ .

B.6 Acceptance sampling methods in which the sampling is conducted in two or more stages are called _____ _____ _____. When using such a method, the results of intermediate samples lead to one of three actions or decisions. Identify these actions. _____

_____

B.7 Refer to the situation in B.1. Suppose the company has decided to switch to the following multiple sampling plan with two stages.

Sample	Sample Size	Cumulative Sample Size	Acceptance Number	Rejection Number
1	25	25	0	2
2	25	50	2	3

Indicate the appropriate action for each of the possible sample outcomes.

a. Sample 1 results in 3 defective parts. _____

b. Sample 1 results in 1 defective part. _____

c. Sample 1 results in 1 defective part and Sample 2 results in 2 additional defective parts. _____

d. Sample 1 results in 1 defective part and Sample 2 results in 0 additional defective parts. _____

B.8 A manufacturer of aluminum arrows inspects incoming lots of N = 15,000 shafts to see if they should be accepted or rejected. The manufacturer uses the following multiple sampling plan.

Sample	Sample Size	Cumulative Sample Size	Acceptance Number	Rejection Number
1	50	50	1	4
2	50	100	3	5
3	50	150	6	7

For a particular lot, suppose Sample 1 results in 3 defective shafts. Describe, in words, the sampling plan to be followed and associated decisions for Sample 2. _____

_____ √
_____

»Problem 2

An assembler of electronic components purchases diodes from a supplier in lots of 50,000. The contract between the two firms specifies an acceptable quality level of three percent defective. The following sampling plan is used to determine the quality level of any particular shipment: Select a random sample of 1,000 diodes from the lot. If 40 or fewer diodes are defective, accept the lot; if 41 or more are defective, reject the lot.

a. State the alternative conclusions for a particular lot in the terminology of a statistical test for p, the proportion of defective diodes in the lot. _____

b. State the decision rule used in this acceptance sampling plan in the terminology of a statistical test for p. _____
   _____ √

c. i. Determine the supplier's risk when p = .03. (Note: From (12.4) and (12.5) in your text, when n is reasonably large (and the population size is large relative to the sample size) the sampling distribution of p for any specified value of p is approximately normal with $E(p) = p = \sqrt{p(1-p)/n}$.) Do not use a correction for continuity in your calculation.

      ii. Would the Supplier's risk have been larger or smaller when p
         = .03 if the acceptance sampling plan had specified an
         acceptance number of 38 rather than 40 diodes? _____
         Explain. _____ √

    d. Determine the buyer's risk when p = .05 given the original
       acceptance number of 40 diodes. _____ √

Answers to Chapter Reviews and Problems

A.1 upper control limit; lower control limit

A.2 time; sample means

A.3 3; α; The α risk measures the probability that it is concluded that
the process mean has changed when, in fact, it has not.

A.4 Since $\bar{X}$ = 1.479 is outside the range specified by the control
limits (i.e., $\bar{X}$ = 1.479 < 1.485), we conclude that the process
mean has changed from its set value of 1.5 cm and, hence, that the
process is out of control.

A.5 If a process is out of control, the process manager will want to
search for the cause of the change. Having found the cause, the
manager may want to alter the causal conditions to bring the
process in control or, if this is not possible or desirable, adjust
the control chart to reflect this change in the process
characteristics.

A.6 Even though the process is still in control, a trend, if it is a
real one, indicates that the process may soon be out of control.
If detected early, it may be possible to adjust the process at
less expense or loss of acceptable output.

A.7 LCL = .03 - 3 $\sqrt{.03(1-.03)/400}$ = .0044;

    UCL = .03 + 3 $\sqrt{.03(1-.03)/400}$ = .0556

A.8 LCL = 8.0 - 3$\sqrt{8.0}$ = -.4853;

    UCL = 8.0 + 3$\sqrt{8.0}$ = 16.49
(Note: Since the number of orders per day that do not receive
"same-day" attention cannot be negative, there is no operational
lower control limit in this situation.)

A.9 $C_1 = (1.503 - 1.5) = .003$;
     $C_2 = .003 + (1.507 - 1.5) = .010$;
     $C_3 = .010 + (1.492 - 1.5) = .002$;
     $C_4 = .002 + (1.504 - 1.5) = .006$

A.10 Since the consecutive values of C_k through Wednesday fluctuated just mildly about 0, it can be concluded that the process remained in control during this period.

Problem 1

   a. H_0: $\mu = .350$ (Let process alone); H_1: $\mu \neq .350$ (Look for cause of change in μ)

   b. $\sigma(\overline{X}) = .015/\sqrt{25} = .003$; Thus, LCL $= .350 - 3(.003) = .341$ and UCL $= .350 + 3(.003) = .359$

   c. (See answer to part (e.i.) of this problem.)

   d. i. (See answer to part (e.i.) of this problem.)

      ii. The process was in control during the three days up until the last sample. It was out of control at 4:00 Wednesday.

   e. i.

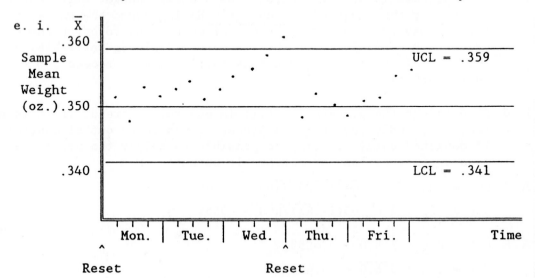

Figure 22.1

ii. Yes; the process was in control during the two days since all
sample means fell between the upper and lower control limits.

f. i.

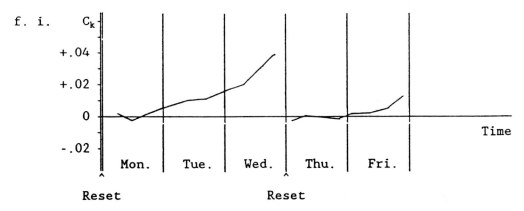

Figure 22.2

ii. Yes. In both instances after the machine was reset, the
cumulative sum at first fluctuated about 0, indicating that
the process was under control. Later, however, the
cumulative sum began a rather steady upward trend, indicating
the process was going out of control. If this pattern
continues, the manager should attempt to determine why the
machine cannot hold settings for a prolonged length of time
and take whatever remedial action is indicated.

B.1 AQL = .03

B.2 2; 3

B.3 acceptable; rejected or sent for 100 percent inspection; producer's
or supplier's

B.4 unacceptable; accepted or sent to inventory; buyer's or purchaser's

B.5 $P(Ho; p)$; operating characteristic curve

B.6 multiple sampling plans; the actions are: reject the lot, accept
the lot or continue sampling.

B.7 a. reject the lot

b. continue sampling

c. reject the lot

d. accept the lot

B.8 Select a second sample of 50 shafts. If no additional defective shafts are found in the second sample, accept the shipment. If one additional defective shaft is found in the second sample (i.e., a total of four shafts in the combined sample of 100 are defective), select a third sample of 50 shafts. If two or more additional defective shafts are found in the second sample, reject the lot.

Problem 2

a. H_0: p ≤ .03 (accept the lot); H_1: p > .03 (reject the lot).

b. If $\bar{p}$ ≤ 40/1000 = .04, conclude H_0 (accept the lot); if $\bar{p}$ > 40/1000 = .04, conclude H_1 (reject the lot).

c. i. $E(\bar{p})$ = .03; $\sigma(\bar{p})$ = $\sqrt{[.03(1-.03)]/1000}$ = .00539
   $P(H_0; p =.03)$ = $P(p \le .04; p =.03)$ = $P[Z\le(.04-.03)/00539]$ = $P(Z \le 1.86)$ = .9686; thus, α = 1 - $P(H_0; p =.03)$ = 1-.9686 = .0314.

   ii. The supplier's risk would have been larger. In this case, we obtain $P(H_0; p=.03)$ = $P(p \le .038; p=.03)$=$P[Z < (.038-.03)/.00539]$ = $P(Z \le 1.48)$ = .9306. Thus, α = 1 - .9306 = .0694.

d. $E(\bar{p})$ = .05; $\sigma(\bar{p})$ = $\sqrt{[.05(1-.05)/1000]}$ = .00689. Thus, β = $P(H_0; p = .05)$ = $P(p \le .04; p = .05)$ = $P(Z \le (.04-.05)/.00689$ = $P(Z < -1.45)$ = .0735.

23

OTHER APPLICATIONS OF SAMPLING

A. SIMPLE RANDOM SAMPLING FROM FINITE POPULATIONS

(Text: Chapter 23, Section 23.1)

Review of Basic Concepts

A.1 When constructing a confidence interval for a population mean μ or a population proportion p with a simple random sample from a finite population when n/N is not small, the appropriate estimated standard deviation includes the term $\sqrt{1 - (n/N)}$, which is called the _____ _____ _____.

A.2 It is desired to estimate the mean number of days of accumulated sick leave among the 2,000 workers in a plant with a simple random sample of 75 workers. Can this population be treated as infinite according to the working rule given in your Text? _____
Explain. _____

A.3 For the situation in A.2, can the population be treated as infinite if the sample size is 300 rather than 75? _____ Explain. _____
_____

A.4 A trucking company wishes to construct a 98 percent confidence interval for μ, with mean age (in miles) of its fleet of 1,600 trucks. A simple random sample of 22<u>5</u> trucks yields X = 47,250 miles and s = 6,500 miles. Hence, s(X) is computed with the formula _____, and equals _____ in this case.

A.5 For the situation in A.4, z(1-α/2) = _____. Thus, the confidence limits are _____ ± _____ and the appropriate confidence interval is _____ $\leq \mu \leq$ _____.

A.6 The finite population correction is nearly equal to _____ when n is small relative to N. Conversely, the finite population correction approaches _____ as n approaches N.

A.7 An alternative sampling procedure to simple random sampling is being considered in a particular situation. It is known that, at the same confidence level and cost, this alternative procedure offers greater precision than simple random sampling. Hence, it is said to be _____ _____ than simple random sampling. √

»Problem 1

An analyst wishes to estimate the mean amount of total assets which are financed by long term debt in a population consisting of 500 companies. He requires a confidence interval with a half-width of $50 thousand and a 90 percent confidence coefficient. Previous research by the analyst suggests that σ = $300 is thousand is a reasonable planning value.

a. i. Determine the required sample size. _____ √

   ii. Based on this sample size, is the finite population correction appropriate in constructing the 90 percent confidence interval? _____ Explain. _____
   _____ √

b. A random sample of n = 82 companies is selected from _this population of 500 companies. The sample results are X = $1310 thousand and s = $315 thousand. Construct a 90 percent confidence interval for μ. _____ √

»Problem 2

a. It is desired to estimate p, the proportion of the 3,000 delegates to a political convention who are in favor of a particular economic plank in the party's platform. In a random sample of n = 600 delegates, 435 were in favor of this plank.

   i. Construct a 99 percent confidence interval for p. _____ √

   ii. Suppose this confidence interval had been mistakenly constructed without the finite population correction. What effect, if any, would this error have had on the width of the confidence interval? _____ Explain. _____
   _____ √

b. A trade association wishes to estimate the proportion of its 2,800 member firms which showed an increase in operating revenue last year. A 95 percent confidence interval with a half-width h = .03 is desired. The planning value p = .75 can be used since p is certain to be no less than .75. Determine the required sample size. _____ √

B. STRATIFIED AND PPS SAMPLING
(Text: Chapter 23, Sections 23.2 and 23.4)

Review of Basic Concepts

B.1 Stratified random sampling is more efficient than simple random sampling whenever it offers _____ precision at the same cost or, alternatively, _____ precision at less cost.

B.2 The population of employees of a firm is divided into two subpopulations by sex of employee for selecting a stratified random sample of employees. The subpopulations are called _____ .

B.3 For the situation in B.2, suppose it was decided to sample n_1 = 150 male employees. Explain by what sampling method these 150 male employees should be selected from the subpopulation of all N_1 = 3,100 male employees. _____

_____

B.4 The community relations officer of a university wishes to select a stratified random sample of alumni to estimate mean current salary of graduates. The officer is attempting to decide whether to use major (e.g., Accounting, History), college (e.g., Business, Arts) or current place of residence to be employed? _____

_____

B.5 For the situation in B.4, suppose the community relations officer wishes to stratify the population of alumni on the basis of both college and current place of residence. Is this possible? _____ Explain. _____

B.6 The total number of strata into which a population is divided is denoted by the symbol _____ .

B.7 Households in a metropolitan area are divided into the following strata on the basis of location of residence within the metropolitan area.

Stratum	Location of Residence	Number of Households	Planning Value for σ_j	Sample Mean Based on Equal Stratum Sample Sizes
1	City	15,000	$220	$770
2	Suburb	22,500	$120	$825

It is desired to estimate mean monthly household spending using a stratified random sample of n - 500 households, equally divided between the strata. Identify each of the following for this study.

a. n_1 = _____ ; n_2 = _____

b. N_1 = _____ ; N_2 = _____ , N = _____

c. $\bar{X}_1$ = _____ ; $\bar{X}_2$ = _____ .

B.8 A point estimator of the population mean when using stratified random sampling is denoted by _____. For the situation in B.7, the point estimate of mean monthly spending is _____.

B.9 For the situation in B.7, what would n_1 and n_2 have been had the sample of n - 500 households been allocated to the strata using proportional allocation?

n_1 = _____ ; n_2 = _____

B.10 For the situation in B.7, what would n_1 and n_2 have been had the sample of n = 500 households been allocated to the strata using optimal allocation?

n_1 = _____ ; n_2 = _____ √

»**Problem 5**

The Department of Employment Services in an eastern region wishes to estimate the mean number of employees among firms in the region. Since it is believed that type of firm is related to number of employees, a stratified random sample is to be used, with type of firm as the basis of stratification. Thus, the population of firms is divided into the subpopulations shown in the following table. Planning values for σ_j were obtained from previously collected data and are believed to be reasonable planning values. A total sample size of n = 1,000 firms is planned.

Stratum j	Type of Firm	N_j	Planning Value for σ_j
1	Retail	42,900	6.5
2	Service	65,000	3.2
3	Manufacturing	9,100	22.5
4	Other	13,000	8.0

a. Determine the sample size for each stratum if proportional allocation is applied.

$n_1 = $ _____ ; $n_2 = $ _____ ;

$n_3 = $ _____ ; $n_4 = $ _____

b. Determine the sample size for each stratum if optimal allocation is applied.

$n_1 = $ _____ ; $n_2 = $ _____ ;

$n_3 = $ _____ ; $n_4 = $ _____ √

c. Explain why the magnitude of n3 differs so greatly between the two methods of allocation.

d. It was decided to use the optimal allocation method and, for convenience, to round the strata sizes to the nearest ten firms. Having done this, the sample was selected and the following results obtained.

Stratum j	N_j	n_j	$\overline{X}_j$	$s_j{}^2$	$(N_j/N)\overline{X}$	$(N_j/N)^2(S_j{}^2/n_j)$
1	42,900	350	17.5	42.0	_____	_____
2	65,000	260	9.0	9.0	_____	_____
3	9,100	260	45.0	450.0	_____	_____
4	13,000	130	32.5	60.0	_____	_____
	130,000	1000		$\overline{X}_{st}$ =	_____	$s\{X\}_{st}$ = _____

i. Obtain $\overline{X}_{st}$ and $s^2\{\overline{X}_{st}\}$ by completing the preceding table. (Note: the finite corrections can be omitted here since the sampling fractions are small.) √

ii. Construct a 90% confidence interval for μ. _____

iii. Interpret this interval. _____
_____ √

C. CLUSTER SAMPLING AND SYSTEMATIC SAMPLING
(Text: Chapter 23, Sections 23.4 and 23.5)

Review of Basic Concepts

C.1 Identify the sampling procedure in each of the following studies.

a. The population of sales representatives in an industry is divided into several geographic regions. Independent simple random samples of sales representatives form each region are then selected.

b. Every 40th name in a computer file of active members of a trade association is selected. The first name is selected at random from among the first 40 names in the file. _____

c. In a study of salaries among university athletic coaches, a simple random sample of universities is selected. The salaries of all athletic coaches employed by each selected university are then determined. _____

d. In a national study of supermarket sales of a new food product, a simple random sample of counties is selected. Within each selected county, supermarkets are classified by size. Independent simple random samples of supermarkets within each size category are then selected. _____

C.2 The sampling procedure in C.1(c.) is single-stage cluster sampling. What comprises a cluster in this situation? _____

_____

C.3 For the situation in C.1(c.), the estimator of the mean salary is likely to be _____ precise than one based on a simple random sample of the same size. This is so because salaries of coaches within the same university are probably _____ homogeneous than in the population of all university coaches. Explain why cluster sampling may nevertheless be more efficient than simple random sampling in this situation. _____

_____

C.4 The sampling procedure in C.1(d.) is multi-stage cluster sampling. In this situation, the primary sampling units are _____ and the secondary sampling units are _____ .

C.5 It is suspected that a carload shipment of a particular brand of disposable razor may not meet company standards regarding product quality. The razors have been packaged 12 to a blister pack. The blister packs have been packed 48 to a carton. There are a total of 720 cartons in this carload. One proposed sampling procedure (Plan A) calls for a simple random sample of 30 cartons, a simple random sample of 6 blister packs within each selected carton, and a measurement of the quality of all 12 razors within each selected blister pack. Thus, the sample size in this study is n = _____ razors, which represents _____ percent of the population.

C.6 For the situation in C.5, an alternative sample procedure (Plan B) is proposed which reduces the number of sampled cartons to 15 and increases the number of sampled blister packs within each carton to 12. Does this proposal alter the sample size? _____ Assuming that the difference in total sampling costs between the two proposals is negligible and that razor quality within each carton is more uniform than between different cartons, which sampling procedure would likely be better? _____ Explain. _____

C.7 A department store has N = 2000 active charge accounts. It is desired to estimate the mean age (i.e., number of years since account was opened) of these accounts with a systematic random sample of every 20th account. The 2000 active accounts are retained in a physical file which is organized alphabetically by name of customer. Describe how this sample is to be selected.

_____

_____

C.8 For the situation in C.7, the resulting sample is called a(n) _____ percent systematic sample. The sample size is n = _____.

C.9 For the situation in C.7, is the estimator of mean account age likely more precise than one based on a simple random sample? _____ Explain. _____

_____ √

Answers to Chapter Reviews and Problems

A.1 finite population correction

A.2 Yes; since the sampling fraction is small (i.e., n/N = 75/2000 = .0375 < .05).

A.3 No; since the sampling fraction is not small (i.e., n/N = 300/2000 = .15 > .05)

A.4 $s\{\bar{X}\} = \sqrt{1 - n/N}\ s/\sqrt{n} = \sqrt{1 - 225/1600}\ (6500/\sqrt{225}) = 401.71$

A.5 $Z(.99) = 2.326;\ 47250 \pm 2.326(401.71);\ 46315.6 \le \mu \le 48184.4$

A.6 1; 0

A.7 more efficient

Problem 1

a. i. $Z(.95) = 1.645;\ \sigma = 300;\ h = 50;\ N = 500$. Thus

$$n = \frac{(1.645)^2(300)^2}{(50)^2 + [(1.645)^2(300)^2/500)]} = 81.53 \approx 82 \text{ companies}$$

   ii. Yes; since the sampling fraction n/N = 82/500 = .164 is
       larger than .05.

b. $\overline{X}$ = 1310; $s(\overline{X})$ = $\sqrt{1 - 82/500}$ $(315/\sqrt{82})$ = 31.806; Z(.95) = 1.645;
   the coefficient limits are 1310 ± 1.645(31.806). Thus, the 90
   percent confidence interval for μ is 1257.7 ≤ μ ≤ 1362.3.

Problem 2

a. i. $\overline{p}$ = 435/600 = .725; $s(\overline{p})$ = $\sqrt{1 - 600/3000}$ $\sqrt{[.725(1-.725)/(600-1)}$
      = .0163; Z(.995) = 2.576; The confidence limits are .725 ±
      2.576(.0163). Thus, the 99 percent confidence interval for p
      is .683 ≤ p ≤ .767.

   ii. Since the finite correction, $\sqrt{1 - 600/3000}$ = .8944 is con-
       siderably less than 1.0, the magnitude of s(p) is reduced by
       including the finite correction. Had it been omitted, s(p)
       would have been mistakenly larger and, hence, the confidence
       interval would have been wider than appropriate for this
       situation.

b.
$$n = \frac{(1.96)^2.75(1-.75)}{(.03)^2 + [(1.96)^2.75(1-.75)]/2800} = 622.4 \approx 623 \text{ firms.}$$

B.1 greater; equal or the same

B.2 strata

B.3 The sample of n_1 = 150 male employees should be selected as a
simple random sample of the N_1 = 3100 male employees, and
independently of the selection of female employees from the other
subpopulation.

B.4 If only one basis of stratification can be used, the officer should
choose the one most closely related to current salary. Thus, each
stratum should be comprised of alumni who are as similar as
possible to one another with respect to current salary (i.e.,
relatively homogeneous) so that salary differences are as great as
possible across strata. Further information would be needed to
determine which basis best satisfies the criterion.

B.5 Yes; provided alumni records cross-reference college and place of
residence for graduates. The officer can then simply divide the
population on the basis of combinations of college and current
place of residence. Thus, the strata would be (Business/instate),
(Business/out-of-state), (Arts/instate), and so on.

B.6 k

B.7 a. $n_1 = 250$; $n_2 = 250$

b. $N_1 = 15000$ $N_2 = 22500$; $N = 37500$

c. $\overline{X}_1 = 770$; $\overline{X}_2 = 825$

B.8 $\overline{X}_{st}$; $\overline{X}_{st} = (15,000/37,000)(700) + (22,500/37,500)(825) = \$803$

B.9 $n_1 = .4(500) = 200$; $n_2 = .6(500) = 300$

B.10 $N_1\sigma_1 + N_2\sigma_2 = 15,000(220) + 22,500(120) = 3,300,000 + 2,700,000$
$= 6.000,000$;
$n_1 = (3,300,000/6,000,000)(500) = 275$;
$n_2 = (2,700,000/6,000,000)(500) = 225$

Problem 3

a. $N = 130,000$ $n_1 = (42,900/130,000)(1,000) = 330$
$n_2 = (65,000/130,000)(1,000) = 500$
$n_3 = (9,100/130,000)(1,000) = 70$
$n_4 = (13,000/130,000)(1,000) = 100$

b.
$$\sum_{j=1}^{4} N_j\sigma_j = 42,000(6.5) + 65,000(3.2) + 9,100(22.5) + 13,000(8.0)$$

$$= 278,850 + 208,000 + 204,750 + 104,000 = 795,600$$

$n_1 = (278,850/795,600)(1,000) = 350.49 \approx 351$;
$n_2 = (208,000/795,600)(1,000) = 261.44 \approx 261$;
$n_3 = (204,750/795,600)(1,000) = 257.35 \approx 257$;
$n_4 = (104,000/795,600)(1,000) = 130.72 \approx 131$

c. Since the proportion of manufacturing firms in the population is small, n_3 is small under proportional allocation. However, since σ_3 is large compared to other planning values for σ_j, $N_3\sigma_3$ is considerably larger relative to $\sum\limits_{j=1}^{4} N_j\sigma_j$ than N_3 is relative to N.

Thus, n_3 is relatively larger under optimal allocation than under proportional allocation.

d. i.

Stratum j	$(N_j/N)\bar{X}_j$	$(N_j/N)^2(s_j^2/n_j)$
1	5.775	.0131
2	4.500	.0087
3	3.150	.0085
4	3.250	.0046

$$\bar{X}_{st} = 16.675 \quad s^2(\bar{X}_{st}) = .0349$$

ii. $s(\bar{X}_{st}) = \sqrt{.0349} = .1868$; $Z(.95) = 1.645$; the confidence limits are $16.675 \pm 1.645(.1868)$. Thus, the 90 percent confidence interval for μ is $16.368 \le \mu \le 16.982$.

iii. With 90 percent confidence, we can state that the mean number of employees per firm in this state is between 16.4 and 17.0.

C.1 a. stratified sampling

b. systematic sampling

c. single-state cluster sampling

d. multi-stage cluster sampling, where the probability sample in the first stage is a simple random sample and in the second stage is a stratified random sample.

C.2 A cluster is comprised of all athletic coaches at a given university.

C.3 less; more; the cluster sample is likely to be less costly. For example, a listing of all universities, needed for cluster sampling, is readily obtained but a frame of all athletic coaches, needed for simple random sampling, would have to be constructed, probably at considerable expense. Also, the actual cost of sampling coaches is probably proportionate to the number of universities covered in the sample. However, once a university is contacted, the cost of determining the salaries of all its coaches is likely to be no greater than the cost of determining the salary of one coach selected in a simple random sample.

C.4 counties; supermarkets

C.5 n = 12(6)(30) = 2160; n/N = 2160/[(12)(48)(720)] = .0052 or 52 percent

C.6 No; n = (12)(12)(15) = 2160. Plan A; since razors within a carton are relatively more homogeneous, Plan A will result in a more precise estimate. Hence, with equal sampling costs, Plan A will be more efficient.

C.7 One of the first 20 accounts is selected at random. Suppose, for example, the random number 07 is chosen. Thus, the first sample account is the 7th account in the alphabetical listing. The sample consists of the 7th, 27th, 47th, . . . , 1987th accounts in the listing.

C.8 100/20 = 5; n = .05(2000) = 100

C.9 No. The frame in this situation stratifies the sample by alphabetical order of customer names. Since this bias of stratification is most likely to be related to account age, the estimator will be no more precise than one based on a simple random sample of accounts.

24

TIME SERIES AND FORECASTING: DESCRIPTIVE METHODS

A. SMOOTHING

(Text: Chapter 24, Sections 24.1 and 24.2)

Review of Basic Concepts

A.1 The observation in period of any time series is denoted by _____ .

A.2 Identify whether or not each of the following is a time series.

   a. Total earnings of each employee in a company last year. _____

   b. Yearly advertising expenditures by the company from 1970 to last year. _____

   c. Daily sales volume for the company last year. _____

A.3 Smoothing is based on the premise that a time series consists of two basic parts, a(n) _____ component and a(n) _____ component. The component which measures relatively short-term and/or erratic movements in the series is the _____ component.

A.4 The purpose of smoothing is to dampen the effect of the _____ component in a time series so that the _____ component can be identified and analyzed.

A.5 The number of terms or observations used to calculate a moving average is denoted by _____ and should be selected to correspond to the _____ of fluctuations in the time series, or a multiple thereof.

A.6 Observations for the first 12 time periods in a time series are 42, 87, 19, 39, 78, 16, 39, 74, 17, 37, 72, and 16. Does the period or cycle of fluctuations in this time series appear to span 2, 3, or 4 time periods? _____ Explain. _____

_____

A.7 For the situation in A.6, if a three-term moving average is used, the first moving average corresponds to t = _____ and is obtained by averaging the three observations_____, _____, and _____ .

A.8 For the situation in A.6, the second moving average, corresponding to t = 3, is obtained by averaging the three observations _____, _____ and _____ .

A.9 A plot over time of the moving averages in a time series can be used to analyze the _____ component of the series. Also, a plot over time of the differences between the observations in the time series and the corresponding moving averages can be used to analyze the _____ component of the series.

A.10 A time series of monthly production volume (in units) of a product is to be smoothed with a 12-term moving average. Explain why the moving averages should be centered. _____
_____
_____

A.11 For the situation in A.10, the first three 12-term moving averages are 525, 625, and 660. Therefore, the first centered 12-term moving average is _____ which corresponds to t = _____ in the time series. √

»Problem 1

The owner of a "supermarket-type" retail toy store is in the process of analyzing quarterly sales for the first five full years of operation, 1982-1986. The time series is to be smoothed in order to better analyze both long-term movements and quarter-to-quarter fluctuations in sales. Quarterly sales data (in $ thousands) appear in Table 24.1, together with some initial computations.

    a. Using Table 24.1, obtain the centered moving averages and fluctuating component of the time series for 1985 and 1986. √

    b. i. Construct a plot of the time series observations and centered moving averages.

      ii. Describe the movement in the smoothed series. _____

      _____

      _____ √

   c. i. Construct a plot of the fluctuating component of the time
       series.

      ii. Describe the quarter-to-quarter fluctuations in the time
       series. _____

      _____

      _____ √

B. CLASSICAL TIMES SERIES MODEL AND LINEAR TREND

(Text: Chapter 24, Sections 24.3 and 24.4)

Review of Basic Concepts

B.1 The model $Y = T * C * S * I$ is called the _____
_____ time series model. Identify the components in this
model. _____

_____

B.2 Describe the pattern or appearance of each of the following
components in a time series as it is plotted against time.

   a. Trend _____

   b. Cyclical _____

   c. Seasonal _____

   d. Irregular _____

Table 24.1

QUARTERLY SALES DATA FOR PERIOD 1982-1886
FOR RETAIL TOY STORE
(See Problem 1)

Year	Quarter	Sales	Moving Total (k=4)	Moving Average (k=4)	Centered Moving Average	Fluctuating Component
1982	1	232				
	2	271				
			1,178	294.50		
	3	252			312.00	-60.00
			1,318	329.50		
	4	423			351.88	71.12
			1,497	374.25		
1983	1	372			392.75	-20.75
			1,645	411.25		
	2	450			447.88	2.12
			1,938	484.50		
	3	400			499.75	-99.75
			2,060	515.00		
	4	716			529.75	186.25
			2,178	544.50		
1984	1	494			559.12	-65.12
			2,295	573.75		
	2	568			589.62	-21.62
			2,422	605.50		
	3	517			598.50	-81.50
			2,366	591.50		
	4	843			583.50	259.50
			2,302	575.50		
1985	1	438			——	——
	2	504	——	——	——	——
	3	418	——	——	——	——
	4	664	——	——	——	——
1986	1	451	——	——	——	——
	2	533	——	——	——	——
	3	539	——	——	——	——
	4	894	——	——		

B.3 The linear trend function is T_t = _____ .   What do
the symbols T_t and X_t in this model denote? _____

_____

B.4 Monthly sales tax receipts (in \$ thousands) in a state during the
period January 1977 through December 1986 have been modeled with a
linear trend function. The coefficients b_0 and b_1 were calculated
by the least squares method using n = _____ observations
(X_t, Y_t), where X_t = 1 at January 1977 and X is in one-month units.

B.5 For the situation in B.4, b_0 = 2,650 and b_1 = 8.75. Interpret the
value b_1 = 8.75. _____

_____

B.6 For the situation in B.4 and B.5, the projected trend value for
January 1987 is _____ .

B.7 A linear trend function was used to model the trend component of a
time series of total annual paid attendance for a minor league
baseball team from 1956 to 1985 where X_t = 1 at 1956, X in one-year
units. The trend value for 1985 was T_{30} = 64,555. Actual paid
attendance in 1985 was Y_{30} = 67,230. The computation
100(67,230/64,555) = _____ is referred to as the
_____ of _____ for 1985 and is a
measure of the _____ component for the year 1985. √

   »Problem 2

    Table 24.2 presents data on the total number of clients served
annually by a small law firm during the period 1980-1986. A linear trend
function is to be used to model the trend component in this time series.

Table 24.2

ANNUAL NUMBER OF CLIENTS SERVED
BY LAW FIRM FOR PERIOD 1980-1986
(See Problem 2)

Year	(1) Coded Year X_t	(2) Number of Clients Y_t	(3) $X_t Y_t$	(4) $X^2{}_t$	(5) Trend Value T_t	(6) Percent of Trend
1980	1	69	69	1	_____	_____
1981	2	78	156	4	_____	_____
1982	3	84	252	9	_____	_____
1983	4	103	412	16	_____	_____
1984	_____	105			_____	_____
1985	_____	133			_____	_____
1986	_____	128			_____	_____
Total	_____	_____	_____	_____		

a. i. Complete the computations in columns (1) through (4) of Table 24.2. √

   ii. Determine the coefficients b_1 and b_0 for the fitted trend equation. b_1 = _____ b_0 = _____

  iii. State the fitted trend equation. _____ √

b. Obtain the trend values and percent of trend values for the period 1980-1986, and insert them in Table 24.2, columns (5) and (6). √

c. i. Plot the fitted trend equation and time series observations.

   ii. Does the linear trend equation reasonably model the trend component in this time series? _____ Explain. _____
      _____ √

d. i. Obtain the projected trend value for 1990. _____

ii. Do you expect that the number of clients in 1990 will exactly equal the projected trend value? _____ Explain. _____
                                                                                  √
_____

C.   EXPONENTIAL TREND
(Text: Chapter 24, Section 24.5)

Review of Basic Concepts

C.1 A trend whose values change by a constant rate or percent from one period to another is called a(n) _____ trend while one whose values change by a constant amount is called a(n) trend.

C.2 An exponential trend appears as a straight line when plotted against a(n) _____ vertical scale.

C.3 An exponential trend can be fitted to a time series by applying the least squares procedure for linear trend to the _____ to base of the Y_t's, which are denoted by _____.

C.4 The exponential trend function is $T_t' =$ _____, where T_t' denotes the _____ of the _____ _____ for period t.

C.5 The first three values in a time series are $Y_1 = 682$, $Y_2 = 720$, and $Y_3 = 737$. The logarithms (to base 10) of these values are $Y_1' = 2.8338$, $Y_2' =$ _____, and $Y_3' =$ _____.

C.6 The following fitted exponential trend equation for a time series was obtained: $T_t' = 2.0175 + .1566X_t$ where $X_t = 1$ at 1980 and X is in one-year units. The logarithm of the trend value in 1985 is _____ and the trend value in 1985 is _____.

C.7 The constant growth rate in an exponential trend is denoted by _____. For the situation in C.6, the growth rate is (antilog .1566)-1 = _____. √

»Problem 3

The Special Chemicals Division of a large corporation has been experiencing declining sales. Sales (in $ millions) for this division for the period 1982-1986 appear in the following table.

Year	Coded Year X_t	Sales Y_t	Y_t'	$X_t Y_t'$	X_t'	Logarithm of Trend Value T_t'	Trend Value T_t
1982	1	35.8	1.5539	1.5539	1	_____	_____
1983	2	28.7	1.4579	2.9158	4	_____	_____
1984	3	23.2	1.3655	4.0965	9	_____	_____
1985	4	21.6				_____	_____
1986	5	20.6	_____	_____	_____	_____	_____
Total	_____		_____	_____	_____		

a. i. Complete the computations in the preceding table needed to determine the coefficients b_1 and b_0 for the fitted exponential trend equation. b_1 = _____ ; b_0 = _____ .

   ii. State the fitted exponential trend equation. _____ . √

b. Determine the trend values T_t for the period 1982-1986 in the preceding table. √

c. i. Construct a plot of the fitted exponential trend equation and time series observations. (Note: To plot the trend equation, first plot the trend values for the years 1982-1986. Then draw a smooth curve through these points.)

   ii. Does the exponential trend equation reasonably model the trend component in this time series? _____ Explain.

d. Determine the value of and interpret the growth rate in this trend curve. _____ √

D. GROWTH-CURVE TREND
(Text: Chapter 24, Section 24.6)

Review of Basic Concepts

D.1 Describe the long-term sweep in a time series with a Gompertz trend. _____

D.2 A time series with a Gompertz trend was analyzed with the following results: $b_0 = 2,000$, $b_1 = .012$, and $b_2 = .85$. Therefore, the Gompertz trend function is _____ .

D.3 For the situation in D.2, the upper asymptote or saturation point of the trend curve is _____ .

D.4 For the situation in D.2, the coefficient $b_2 = .85$ is called the _____ coefficient. What is reflected by this coefficient?

_____

_____

D.5 A time series with a logistic trend was analyzed with the following results: $b_0 = 8,000$, $b_1 = 15.6$ and $b_2 = .70$. Therefore, the logistic trend function is _____ .

D.6 For the situation in D.5, annual time series data were coded so that $X_t = 1$ at 1976. Thus, the trend value for 1985 is _____ . ✓

»Problem 4

A research analyst for a packaged foods manufacturer has fit a Gompertz trend function to annual world-wide sales volume (in millions of units) of a particular cake mix. Years were coded so that $X_t = 1$ at 1966, the year the product was introduced, and X is in one-year units, Based on this analysis, it was found that $b_0 = 400$, $b_1 = .123$, and $b_2 = .80$.

a. State the Gompertz trend function. _____ ✓

b. Determine the trend value for the following selected years. (Note: Higher order exponents must be computed first. For example, 4^{3^2} is computed as 4 raised to the $3^2 = 9$th power, or 4^9).

Year	X_t	T_t	Year	X_t	T_t
1966	1	74.8	1971	____	____
1967	2	104.6	1972	____	____
1968	3	136.8	1975	____	____
1969	4	169.5	1980	____	____
1970	____	____	1985	____	____

c. Using the trend values from (b.), describe the long-term sweep
in this time series. _____

_____

_____

d. Identify and explain the meaning of the upper asymptote in this
growth curve. _____

_____ √

E. FORECASTING BY TREND PROJECTIONS, CYCLICAL ANALYSIS AND SHORT-TERM FORECASTING

(Text: Chapter 24, Sections 24.7 and 24.8)

Review of Basic Concepts

E.1 A trend curve for an annual time series is useful in short-term
forecasting only if the _____ component is a minor
influence on the time series.

E.2 A manufacturing firm has developed a trend equation which is
utilized to project trend values for annual shipments by the firm
of decorative molding used in domestic automobile assembly.
Identify whether or not the trend projections would likely have to
be modified under each of the following potential new conditions.

a. A change in direction of consumer preferences for domestic
versus imported automobiles. _____

b. The introduction of a new technologically superior molding by a
competitor. _____

c. A change in the seasonal fluctuation in shipments caused by
increased emphasis on mid-year model changes in the automobile
industry. _____

E.3 Why is cyclical analysis usually very difficult when dealing with
business and economic time series? _____

_____

E.4 The point in time at which the cyclical component of a time series
changes direction is called a cyclical _____.

E.5 Quarterly sales (in $ millions) of a manufacturer of motorized recreational vehicles for the period 1974-1986 are shown in the following graph.

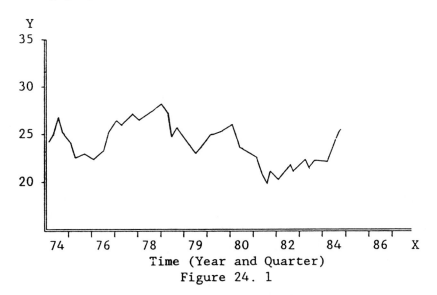

Figure 24. 1

Identify periods in which cyclical downturns in this time series appear to begin.

E.6 It is generally quite hard to identify a current cyclical turning point in a time series both because of the changing duration and amplitude of the cyclical behavior and also because of the presence of _____ fluctuations in the time series.

E.7 The following news item was reported on a TV newscast: "The Government's Index of Leading Indicators declined by almost one percent last month. This is the second consecutive month in which the Index declined." Interpret the possible implication of this decline. _____

_____

E.8 Is it reasonable that the time series "Index of New Private Housing Units Authorized by Local Building Permits" is a leading indicator of U.S. business and economic activity? _____ Explain. _____

_____

E.9 Cyclical indicators which help identify cyclical turning points during and after their occurrence are called, respectively, _____ and _____ indicators. √

F.   CALCULATION AND APPLICATIONS
OF SEASONAL INDEXES
(Text: Chapter 24, Sections 24.9 and 24.10)

Review of Basic Concepts

F.1 In the classical multiplicative time series model, the seasonal component describes repetitive _____ movements in the time series of duration _____ _____ or less.

F.2 In the method of ratio to moving average, the moving averages represent the _____ component in the time series. Thus, if Y observations are divided by corresponding moving averages, the results represent the _____ component which, expressed as percents, are called _____ _____.

F.3 One observation in a time series is 120. The moving average corresponding to this observation is 135. Thus, the corresponding specific seasonal relative is _____.

F.4 The medians of the specific seasonal relatives in a monthly time series of sales of garden supplies by a hardware wholesaler sum to 1,250. To obtain the appropriate seasonal indexes, each median should be multiplied by the factor _____. Thus, if the median of the specific seasonal relatives for January is 95.0, the seasonal index for January is _____. Interpret this seasonal index. _____

F.5 For the situation in F.4, sales last January were $158,916. Seasonally adjusted sales last January were, then, _____. Also, the seasonally adjusted annual rate of sales last January was _____.

F.6 Seasonally adjusted values for a time series obtained by dividing the observations by their respective seasonal indexes represent the _____ component of the series.

F.7 In a time series of housing starts, the observation for May 1986 is 9,200. The trend value for this period is 8,000. Hence, the percent of trend is _____, which represents the _____ component of the series for May, 1986. The seasonal index for May is 125. Thus, the deseasonalized percent of trend for May 1986 is _____, which represents the _____ component of the series for May 1986.

F.8 The forecast for total sales in a particular industry next year is $66 billion, with trend-cyclical movements within the year expected to be minor. Seasonal indexes for January, February, and March are 65, 72, and 112, respectively. Forecasted industry sales in January, February, and March are, therefore, _____, _____ and _____. √

»Problem 5

In Problem 1 we used a smoothing procedure to separate the systematic And fluctuating components of quarterly sales (in $ thousands) of a retail Toy store during the period 1982-1986. Let us reanalyze that time series using the classical multiplicative time series model. Relevant data from Problem 1 appears in Table 24.1. Specific seasonal relatives for the years 1982, 1983, and 1984 have been obtained from the data in Table 24.1 and appear as follows:

	Quarter				
Year	1	2	3	4	Total
1982			80.77	120.21	
1983	94.72	100.47	80.04	135.16	
1984	88.35	96.33	86.38	144.47	
1985	____	____	____	____	
1986	____	____	____	____	
Median	____	____	____	____	_____
Seasonal Index	____	____	____	____	_____

a. i. Obtain the specific seasonal relatives for 1985 and 1986 and insert them in the preceding table. √

ii. Determine the median specific seasonal relative for each quarter. (Note: When the number of values is even, the median is defined by order of magnitude.)

iii. Obtain the seasonal indexes for this time series by adjusting the medians of the specific seasonal relatives. √

b. i. Obtain the seasonally adjusted quarterly sales for the toy store for each quarter of 1986.

Quarter:	1	2	3	4
Seasonally Adjusted Sales:	_____	_____	_____	_____

ii. What time series components are included in the deseasonalized quarterly sales for 1986? _____. √

c. The owner of the toy store has just forecast annual sales of $2,800 thousand for next year. Using the seasonal indexes developed in (a.), obtain forecasts of sales by quarter for next year, assuming that trend-cyclical movements within the year are expected to be minor. √

Quarter:	1	2	3	4
Forecast:	_____	_____	_____	_____

Answers to Chapter Reviews and Problems

A.1 Y_t

A.2 a. No; b. Yes; c. Yes

A.3 systematic; fluctuating; fluctuating

A.4 fluctuating; systematic

A.5 k; period or cycle

A.6 3; if the series is organized into groups of three periods (i.e., [42, 87, 19], [39, 78, 16], [39, 74, 17] and [37, 72, 16]), a pronounced repetitive pattern emerges. In each group of three periods, the series increased from the first to the second observation, then decreases sharply from the second to the third observation.

A.9 t = 2; 42, 87 and 19

A.8 87, 19, and 39

A.9 systematic; fluctuating

A.10 Since the number of terms (12) is even, the moving averages will be off-center. For example, the first moving average will fall midway between the sixth and seventh observation. This problem is corrected by centering the moving averages.

A.11 (525 + 625)/2 = 575; t = 7

Problem 1

a.

Year	Quarter	Sales	Moving Total (k=4)	Moving Average (k=4)	Centered Moving Average	Fluctuating Component
1985	1	438			563.12	-125.12
			2,203	550.75		
	2	504			528.38	-24.38
			2,024	506.00		
	3	418			507.62	-89.62
			2,037	509.25		
	4	664			512.88	151.12
			2,066	516.50		
1986	1	451			531.62	-80.62
			2,187	546.75		
	2	533			575.50	-42.50
			2,417	604.25		
	3	539				
	4	894				

b. i.

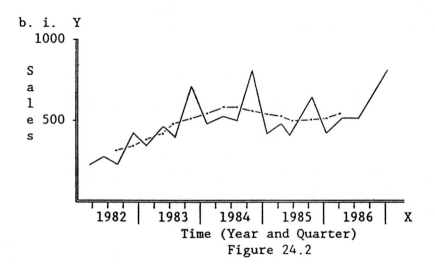

Figure 24.2

    ii. As shown by the plot of the centered moving averages, the
        systematic component of sales increased steadily from 1982
        until the third quarter of 1984. It then turned downward
        until the third quarter of 1985. After that, the systematic
        component increased again. The moving averages, of course, do
        not indicate why these turning points occurred. Further
        analysis would be needed to indicate whether they resulted
        from changes in consumer income levels, management policy
        changes, or other possible factors.

c. i.

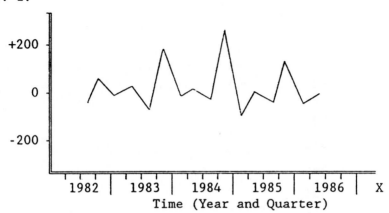

Figure 24.3

    ii. Sales volume appears to follow a repetitive four-quarter
        period of fluctuations. There is a tendency for sales to
        increase slightly from the first to the second quarter and to
        decrease about the same amount from the second to the third
        quarter of each year. Sales then rise very sharply to a
        fourth quarter peak before, once again, falling from the
        fourth quarter to the first quarter. The fourth quarter peak
        in sales likely results from increased consumer toy buying
        during the holiday season.

B.1 classical multiplicative; the four components in this model are
trend (T), cyclical (C) , seasonal (S) , and irregular (I).

B.2 a. Trend component appears as a smooth curve spanning the entire
    time series.

b. Cyclical component appears as cycles which vary both in amplitude and duration.

c. Seasonal component appears as a sequence of relatively repetitious short cycles.

d. Irregular component appears as an irregular, saw-tooth shaped pattern.

B.3 $T_t = b_0 + b_1X_t$; T_t denotes the trend value in any given time period t and X_t is a numerical code representing time period t.

B.4 n = 120

B.5 The trend component in the time series of sales tax receipts increases by $8.75 thousand each month.

B.6 $T_{121} = 2650 + 8.75(121) = \$3,708.75$ thousand.

B.7 104.1; percent of trend; C·I or cyclical-irregular

Problem 2

a. i.

	(1)	(2)	(3)	(4)
Year	X_t	Y_t	X_tY_t	X_t^2
1984	S	105	525	25
1985	6	133	798	36
1986	7	128	896	49
Total	28	700	3,108	140

ii. $b_1 = \dfrac{3108 - [28(700)]/7}{140 - (28^2/7)} = 11.0$; $b_0 = \dfrac{1}{7}[700 - 11(28)] = 56$

iii. $T_t = 56.0 + 11.0X_t$, where $X_t = 1$ at 1980 and X is in one-year units.

b.

X_t	T_t	Percent of Trend	X_t	T_t	Percent of Trend
1	67	103.0	5	111	94.6
2	78	100.0	6	122	109.0
3	89	94.4	7	133	96.2
4	100	103.0			

c. i.

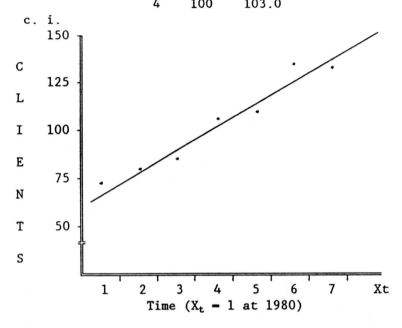

Figure 24.4

    ii. Yes; it appears to follow the long-term sweep of the series.

d. i. $T_{11} = 56.0 + 11.0(11) = 177$

    ii. No, it is unlikely that the number of clients in 1990 will exactly equal 177. Even if the linear trend function continues to be appropriate, cyclical and irregular influences will likely affect the actual number of clients in 1990.

C.1 exponential; linear

C.2 ratio or logarithmic

C.3 logarithms; 10; Y_t'

C.4 $T' = b_0 + b_1 X_t$; logarithm, trend value

C.5 $Y_2' = 2.8573$; $Y_3' = 2.8675$

C.6 $T_6' = 2.0175 + .1566(6) = 2.9571$; $T_6 = $ antilog $2.9571 = 905.9$

C.7 g; $1.434 - 1 = .434$, or 43.4 percent per year

Problem 3

a. i.

Year	X_t	Y_t	Y_t'	$X_t Y_t'$	$X_t{}^2$
1985	4	21.6	1.3345	5.3380	16
1986	5	20.6	1.3139	6.5695	25
Total	15		7.0257	20.4737	55

$$\text{Thus, } b_1 = \frac{20.4737 - [15(7.0257)]}{55 - 15^2(5)} = -.06034$$

and $b_0 = (1/5)[7.0257 - (-.06034)(15)] = 1.58616$.

ii. $T_t' = 1.58616 - .06034 X_t$ where $X_t = 1$ at 1982 and X is in one year units.

b.

X_t	T_t'	T_t=antilogT_t'
1	1.5258	33.6
2	1.4655	29.2
3	1.4051	25.4
4	1.3448	22.1
5	1.2845	19.3

c. i.

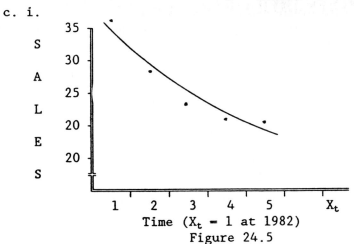

Time ($X_t = 1$ at 1982)

Figure 24.5

ii. Yes; it appears to follow the long-term sweep of the series.

d. $g = [\text{antilog}(-.06034] - 1 = .870 - 1 = -.130$, or -13.0 percent per year. Thus, the trend component in this time series is declining (since g is negative) at the rate of 13.0 percent per year.

D.1 The trend component increases rapidly in the early segment of the time series and then slows down until it eventually approaches a saturation point. In contrast, the rate of growth continually decreases throughout the time series. Plotted on an arithmetic grid, the Gompertz trend appears as an elongated S-shaped curve.

D.2 $T_t - 2,000(.012).85X_t$

D.3 $b_0 - 2,000$

D.4 Retardation; this coefficient reflects the extent or degree of retardation in the rate of growth in the trend component between successive periods.

D.5 $T_t = 8000/[1 + 15.6(.70)X_t$

D.6 $T_{10} = 5553$

Problem 4

a. $T_t = 400(.123).80X_t$

b. The trend value for 1970 is $T_5 - 400(.123)-805 = 400(.123).328 = 400(.503) = 201.2$. Continuing in this fashion, we obtain the remaining trend values.

Year	X_t	T_t
1971	6	230.9
1972	7	257.8
1975	10	319.4
1980	15	371.6
1985	20	390.5

c. The trend component of sales volume increased very rapidly in the years immediately following introduction of the product. However, beginning in 1970 the annual increase began to slow down and, eventually, to level off. For example, the trend value increased just 390.5 - 371.6 = 18.9 million units during the five-year period 1980-1985 compared to an increase of 104.6 - 74.8 = 29.8 million units between 1966 and 1967.

d. The upper asymptote is b_0 = 400. Thus, 400 million units is the upper limit of the trend component in this time series.

E.1 C·I or cyclical-irregular

E.2 a. Yes; b.Yes; c.No

E.3 Cyclical movements in such time series vary a great deal in duration and amplitude, making them difficult to discern in historical time series and even more difficult to predict.

E.4 turning point

E.5 There appear to be three cyclical downturns in the time series; one in the second half of 1974, one in the first half of 1979, and one in the first half of 1982.

E.6 irregular

E.7 It is possible that a cyclical downturn in aggregate economic activity has occurred or is developing.

E.8 Yes; most often, applications for building permits are made prior to the start of construction. Thus, if the number of building permits changes, it usually means that a change in the level of construction activity, with corresponding direct and indirect impact on overall business and economic activity, will occur in the near future.

E.9 coincident; lagging

F.1 cyclical; one year

F.2 T·C or trend-cyclical; S·I or seasonal-irregular; specific seasonal relatives

F.3 100(120/135) = 88.9

F.4 1200/1250 = .96; .96(95.0) = 91.2; sales in January are about 91.2 percent as great as they would be if no seasonal effects were present.

F.5 158916/.912 = $174,250; 12(174,250) = $2,091,000

F.6 T·C·J or trend-cyclical-irregular

F.7 100(9,200/8,000) = 115; C·S·I or cyclical-seasonal-irregular; 115/1.25 = 92; C·I or cyclical-irregular

F.8 .65(66/12) = $3.575 billion; $3.96 billion; $6.16 billion

Problem 5

a. i. and ii.

	Quarter				
Year	1	2	3	4	Total
1982			80.77	120.21	
1983	94.72	100.47	80.04	135.16	
1984	88.35	96.33	86.38	144.47	
1985	77.78	95.39	82.35	12g.46	
1986	84.84	92.62			
Median	86.60	95.86	81.56	132.31	396.33

The median for the first quarter is computed as (84.84 + 88.35)/2 = 86.60. The other three medians are computed in a similar fashion.

iii. Since the medians sum to 396.33, the seasonal indexes are obtained by multiplying each median by the factor 400/396.33. For example, the seasonal index for the first quarter is computed as (400/396.33)86.60 = 87.4. Thus, the seasonal indexes are as follows:

Quarter:	1	2	3	4
Seasonal Index:	87.4	96.8	82.3	133.5

b. i.

Quarter:	1	2	3	4
Seasonally Adjusted Sales:	$\frac{451}{.874}$ = 516;	$\frac{533}{.968}$ = 551;	$\frac{539}{.873}$ = 655;	$\frac{894}{1.335}$ = 670

ii. The seasonally adjusted or deseasonalized quarterly sales for 1986 represent the trend-cyclical-irregular component of the time series.

c.

Quarter:	1	2	3	4
Forecast:	.874(2800/4)−611.8	677.6	576.1	934.5

25

TIME SERIES AND FORECASTING II: SMOOTHING AND REGRESSION

A. EXPONENTIAL SMOOTHING FOR STATIONARY TIME SERIES
(Text: Chapter 2, Section 25.1)

A.1 A stationary time series is one that contains just a(n) _____ component.

A.2 In the exponential smoothing model, the actual or observed value of the time series for period t is denoted by _____, and the smoothed estimate of the mean level of the stationary time series for period t is denoted by _____.

A.3 In a particular situation involving a stationary time series, the smoothed estimate for period 5 was 550 while the value of the time series for period 6 was 510. If a = .25, the smoothed estimate for period 6, which is denoted by _____, equals _____.

A.4 For the situation in A.3, forecasts for the next two periods (periods 7 and 8) are desired. These forecasts, which are denoted by _____ and _____, respectively, equal _____ and _____.

A.5 To initialize or begin smoothing a stationary process, a starting value A_0 is required. Explain how A_0 is obtained. _____
_____

A.6 In a particular situation involving exponential smoothing, a, which is called the _____ _____, has been set at .20. However, management noticed that the forecasting procedure did not respond rapidly enough to occasional changes in the mean level of the time series. This problem can be alleviated by _____ the value of a.

A.7 If the smoothed estimates, A_t, for a stationary time series tend to be overly sensitive to irregular movements in the time series observations, the value of a should be _____.

A.8 In an exponential smoothing problem, the starting value and smoothing constant are specified as follows: $A_0 = 600$ and $a = .20$. Further, the smoothed estimate for period 2 is 601.20 and the time series observation for period 3 is $Y_3 = 615$. Using (25.2) in your Text, the smoothed estimate for period 3 is $A_3 = $ _____.

A.9 Although the smoothed estimate for period t in a stationary time series, A_t, requires explicit knowledge of the time series observation just for period _____, it is actually based on the time series observations for _____ preceding periods.

A.10 For the situation in A.8, the first two observations of the stationary time series were $Y_1 = 620$ and $Y_2 = 590$. Using (25.4) in your Text to show that the smoothed estimate for period 3, A_3, is based on the magnitude of the time series observations for periods 1, 2, and 3 (as well as the starting value $A_0 = 600$), we obtain $A_3 = $ _____.

A.11 For the situation in A.8 and A.10, the smoothed estimate A_3 is a _____ _____ of the starting value $A_0 = 600$ and observations Y_1, Y_2 and Y_3, where the weights for observations Y_1, Y_2 and Y_3 are _____, _____ and _____, respectively, and the weight for the starting value A_0 is _____.

A.12 In an exponential smoothing problem, the smoothing constant has been specified as $a = .10$. Thus, in determining the smoothed estimate for period 10, A_{10}, the time series observation for period 10 has weight _____, the time series observation for period 6 has weight _____, and the time series observation for period 1 has weight _____. ✓

»Problem 1

An inventory manager for a manufacturing company is charged with controlling the inventory of more than 900 items used as fabricating parts and materials. For most of these items, the manager uses exponential smoothing to forecast demand one month ahead. Monthly demand for part #060481 is forecast with an exponential smoothing model for stationary time series. The smoothed estimate for June (period 114) was 2,440, and actual quantity demanded for this part in July (period 115) was 2,415.

a. i. Obtain the smoothed estimate for July and prepare a forecast for demand in August (period 116), if the smoothing constant a = .20 is used. _____ √

ii. Actual demand in August was 2,420. What information must be carried over from July to August to prepare the September forecast? _____
_____

iii. What is forecasted demand for September? _____ √

b. i. If a had been .3 rather than .2, what would your forecasts for August and September have been?
F_{116} = _____ F_{117} = _____

ii. Which of the two values for a would be more appropriate if the manager wishes to place relatively heavy weight on demand in recent months when forecasting future demand? _____ √

B. EXPONENTIAL SMOOTHING FOR TIME SERIES WITH TREND COMPONENT, OR WITH TREND AND SEASONAL COMPONENTS
(Text: Chapter 25, Sections 25.2 and 25.3

Review of Basic Concepts

B.1 In smoothing a time series with trend component, the smoothed estimate for period t is denoted by _____ and the trend estimate for period t is denoted by _____. This latter estimate represents the change in the _____ _____ of the time series at period t.

B.2 The constant used to update the trend estimate in smoothing a time series with trend component is called the _____
_____ constant, and is denoted by _____.

B.3 In a particular situation involving a time series with trend component, the smoothed estimate for period 19 is 4,521.80, the trend estimate for period 19 is 14.30 and the time series observation for period 20 is 4,538.40. If the smoothing constant is .10 and the trend adjustment constant is .25, the updated smoothed estimate for period 20 is _____.

B.4 For the situation in B.3, the updated trend estimate is _____.

B.5 For the situation in B.3 and B.4, it is desired to use the updated smoothed and trend estimates to obtain forecasts for the series in periods 21, 22 and 23. The resulting forecasts are $F_{21} = $ _____, $F_{22} = $ _____, and $F_{23} = $ _____.

B.6 When applying exponential smoothing to a quarterly time series with trend and seasonal components, starting values are needed for the _____ _____ (denoted by A_0), the _____ _____ (denoted by B_0), and the _____ _____ (denoted by C_{-3}, C_{-2}, C_{-1}, and C_0).

B.7 In smoothing a time series with trend and seasonal components, a seasonal index adjustment constant, denoted by ___, is required. Since each seasonal index estimate is updated just once during a seasonal cycle (and, hence, has fewer opportunities to adjust to changes in the seasonal pattern), it may be necessary to use a seasonal index adjustment constant which is relatively _____ compared to the smoothing constant and the trend adjustment constant.

B.8 In a particular situation involving a quarterly time series with trend and seasonal components, $A_{23} = 643.5$, $B_{23} = -8.70$, and the most recently updated seasonal index estimates are $C_{20} = 1.357$, $C_{21} = .845$, $C_{22} = .812$ and $C_{23} = .982$. The appropriate weighting constants are $a = .05$, $b = .20$ and $c = .45$. The current observation for this time series is $Y_{24} = 858.8$. Using (25.7a) in your Text, the updated smoothed estimate for period 24 is _____.

B.9 For the situation in B.8, using (25.7b) in your Text, the updated trend estimate for period 24 is _____, and using (25.7c) in your Text, the updated seasonal index estimate for period 24 is _____.

B.10 For the situation in B.8 and B.9, it is desired to use the most recently updated estimates to forecast the series for the next four quarters. The resulting forecasts are $F_{25} = $ _____, $F_{26} = $ _____, $F_{27} = $ _____ and $F_{28} = $ _____ √

»Problem 2

An analyst for an agribusiness association is responsible for forecasting a variety of time series concerning the consumption of foodstuffs. One such time series, annual per capita consumption (in kilograms) of meat other than red meats in one geographic region, was determined to contain a trend component. Observations on this series between 1971 and 1986 are shown in Table 25.1, where t - 1 at 1971. Starting valleys for the smoothed and trend estimates, obtained by visually examining a plot of the time series observations over time, were A_0 - 30.0 and B_0 - 1.2. Weighting constants were specified as a = .15 and b - .25.

Table 25.1
PER CAPITA CONSUMPTION OF MEAT OTHER THAN RED
MEATS FOR THE PERIOD 1971-1986
(See Problem 2)

Yr. t	Per Capita Consumption Y_t	Smoothed Estimate A_t	Trend Estimate B_t	Yr. t	Per Capita Consumption Y_t	Smoothed Estimate A_t	Trend Estimate B_t
1	30.9	31.155	1.1888	9	41.7	41.073	1.2108
2	32.6	32.382	1.1984	10	42.3	42.286	1.2114
3	34.6	33.733	1.2366	11	43.8	43.543	1.2228
4	36.0	35.125	1.2752	12	45.5	44.876	1.2503
5	36.6	36.430	1.2827	13	45.3	46.002	1.2193
6	37.0	37.806	1.2560	14	46.4	_____	_____
7	37.6	38.672	1.2087	15	48.9	_____	_____
8	39.2	39.779	1.1831	16	49.5	_____	_____

a. Complete Table 25.1 by obtaining smoothed and trend estimates for periods 14, 15, and 16. √

b. i. Construct a graph showing the time series observations and smoothed estimates for the period 1971-1986.

   ii. Based on your graph in part (ii), does it appear that the smoothing procedure used here would provide reasonable forecasts for future periods of the time series? - _____
   Explain. _____

c. Using your results in part (i.), forecast annual per capita consumption of meat other than red meats in this region for 1987 to 1990, inclusive.

1987: _____   1989: _____

1988: _____   1990: _____ √

»Problem 3

An administrator with a federally affiliated student loan program uses exponential smoothing to forecast numerous time series concerning student enrollment patterns. One such series, determined to contain trend and seasonal components, tracks enrollment (in thousands) at all private, four-year colleges in one state by term fall semester, spring semester, summer sessions. Table 25.2 presents this series by term between 1980 and 1986, where t - 1 at f.11 semester, 1980. Starting values for this series were A_0 - 70.5, B_0 - -1.1, C_{-2} - 1.30, C_{-1} - 1.20 and C_0 = 0.50. Weighting constants are a = .15, b = .20 and c = .40.

Table   25.2

ENROLLMENT (in thousands) AT ALL PRIVATE FOUR-YEAR
COLLEGES IN ONE STATE, BY TERM
(See Problem 3)

Year	and Term	Period t	Enrollment Y_t	Smoothed Estimate A_t	Trend Estimate B_t	Seasonal Index Estimate C_t
1980	Fall	1	91.2	69.513	-1.0774	1.3048
	Spring	2	80.3	68.208	-1.1234	1.1909
	Summer	3	34.9	67.492	.1.0419	0.5068
1981	Fall	4	86.5	66.427	-1.0466	1.3038
	Spring	5	78.9	65.511	-1.0204	1.1963
	Summer	6	32.0	64.288	-1.0609	0.5032
1982	Fall	7	81.7	63.143	-1.0778	1.2998
	Spring	8	74.8	62.134	-1.0639	1.1993
	Summer	9	32.2	61.509	-0.9763	0.5113
1983	Fall	10	78.1	60.465	-0.9897	1.2965
	Spring	11	70.0	59.309	-1.0229	1.1917
	Summer	12	29.9	58.315	-1.0172	0.5119
1984	Fall	13	72.5	57.091	-1.0586	1.2859
	Spring	14	67.7	56.149	-1.0353	1.1973
	Summer	15	28.3	55.140	-1.0301	0.5124
1985	Fall	16	72.0	54.392	-0.9736	1.3010
	Spring	17	64.3	53.461	-0.9650	1.1995
	Summer	18	27.8	52.760	-0.9124	0.5182
1986	Fall	19	65.5	_____	_____	_____
	Spring	20	57.9	_____	_____	_____
	Summer	21	26.0	_____	_____	_____

a. Complete Table 25.2 by obtaining smoothed, trend and seasonal index estimates for 1986. √

b. Using your results in part (a.), forecast enrollment at all private four-year colleges in this state by term for 1987.
   Fall Semester: _____
   Spring Semester: _____
   Summer Sessions: _____ √

C. REGRESSION MODELS WITH INDEPENDENT ERROR TERMS
(Text: Chapter 25, Section 25.4

Review of Basic Concepts

C.1 The basic multiple regression model for time series, model (25.9) in your Text, incorporates time-related effects into the model through the _____ _____ and assumes that the error terms ϵt are_____ of one another and are _____ distributed with a mean of _____ and _____ variance.

C.2 State the regression model which is analogous to a linear trend function. _____

C.3 Regression model (25.11) in your Text was used to analyze quarterly sales (in $ thousands) of a firm for the past fifteen years. The estimated response function was $\hat{Y}_t = 420 + 4.3X_{t1} + 2.3X_{t2} + 1.6X_{t3} - 0.7_{X4}$, where $X_{t1} = t$ and independent variables X_{t2}, X_{t3}, and X_{t4} are indicator variables for differential seasonal effects in the second, third, and fourth quarters, respectively. A point estimate of the change in trend level from one quarter to the next is _____. Interpret this result. _____

C.4 For the situation in C.3, the indicator variables are coded so that all are zero for the first quarter. An estimate of the differential seasonal effect in the fourth quarter as compared to that in the first quarter is _____.

C.5 Independent or predictor variables which refer to the same time period as the dependent variable are called _____ variables while those which refer to earlier time periods are called _____ variables.

C.6 The linear trend regression model with independent error terms, model (25.10) in your Text, was used to analyze annual directory assistance calls at a local telephone office for the past 20 years, with these results: $\hat{Y}_t = 3255 + 185X_t$, where $X_t = t$. Also, $s(b_1) = 6.2$. With 90 percent confidence we can state that the expected number of calls increases by between _____ and _____ each year.

C.7 A multiple regression model with independent error terms is to be developed to analyze quarterly retail sales of building materials in a particular county. In addition to a linear trend and a seasonal component, the model is to include the number of building permits issued in the preceding quarter and disposable earnings in the county in each of the preceding two quarters. State the appropriate model for this analysis. _____

C.8 For the situation in C.7, disposable earnings represents a variable with a(n) _____ _____ _____ in the model. √

»Problem 4

An analyst for a local transit commission utilized the linear trend model (25.10) in your Text to study the time series of annual ridership (in millions of riders) during the period 1977-1986. The estimated regression function is $\hat{Y}_t = 2.341 + .181X_t$. Other relevant results from this analysis appear in Table 25.3.

Table 25.3
RESULTS OF ANALYSIS OF TRANSIT
RIDERSHIP FOR PERIOD 1977-1986
(See Problem 4)

Year	Coded Year X_t	Number of Riders (millions) Y_t	$\hat{Y}_t$	Residual $Y_t - \hat{Y}_t$
1977	1	2.48	2.522	-.042
1978	2	2.78	2.703	.077
1979	3	2.85	2.884	-.034
1980	4	3.09	3.065	.025
1981	5	3.29	3.246	.044
1982	6	3.38	3.427	-.047
1983	7	3.57	_____	_____
1984	8	3.80	_____	_____
1985	9	3.90	_____	_____
1986	10	4.23	_____	_____

a. i. Complete Table 25.3 by obtaining $\hat{Y}_i$ and the residuals for the years 1983 through 1986. √

  ii. Plot the residuals by time order.

 iii. Based on the residual plot, does the assumption of independent error terms appear warranted? _____ Explain.
_____ √

b. i. Given $s(b_1) = .0063$, construct a 90 percent confidence interval for β_1. _____.

  ii. Interpret this confidence interval. _____
_____ √

c. Given $s(Y_{h(new)}) = .070$ when $X_h = 11$, construct a 90 percent prediction interval for annual ridership in 1987. _____ √

»Problem 5

An economist used the following multiple regression model to analyze quarterly shipments of lumber in a region for the period 1974-1986.

$Y_t = \beta_0 + \beta_1 X_{t1} + \beta_2 X_{t2} + \beta_3 X_{t3} + \beta_4 X_{t4} + \beta_5 X_{t-4,5} + \epsilon_t$, where
Y_t = quarterly shipments (in S millions) in the region
$X_{t1} = t$
X_{t2}, X_{t3}, and X_{t4} are indicator variables for seasonal effects in quarters 2, 3, and 4, respectively.
$X_{t-4,5}$ is disposable personal income (in \$ billions) in the region lagged 4 quarters
ϵ_t's are independent $N(0,\sigma^2)$

Computer output from the analysis of this regression model appears in Table 25.4.

Table 25.4
COMPUTER OUTPUT FOR REGRESSION
ANALYSIS OF QUARTERLY SHIPMENTS
OF LUMBER FOR PERIOD 1974-1986
(See Problem 5)

Variable	Reg. Coef.	Std. Dev.	T Stat.
Constant	11.712	2.502	4.68
X1	.383	.047	8.15
X2	6.770	1.563	4.33
X3	3.245	1.040	3.12
X4	-.423	.572	-.74
X5	.825	.212	3.89

Analysis of Variance Table

Source	SS	df	MS
Regression	2,213.50	5	442.70
Residual	624.22	46	13.57
Total	2,837.72	51	

Variable	Specified Level
X1	54
X2	1
X3	0
X4	0
X5	31.6

Y Estimate = 65.234 Std. Dev. Mean Response = 4.47
Std. Dev. New Response = 5.79

a. Conduct the appropriate F test with α = .01 to determine whether or not quarterly sales are related to the set of independent variables as specified in the preceding model. Specify the alternatives, determine F* and the decision rule and state the resulting decision. (Note: F(.99; 5, 46) = 3.45). √

b. State the estimated regression function. _____

c. i. Obtain the value of the estimated regression function for the second quarter of 1987 (t=54), given that disposable personal income in the second quarter of 1986 was $X_{50,5}$ = 31.6 and that the indicator variables for seasonal effects are coded X_{t2} = 1, X_{t3} = 0, and X_{t4} = 0 in the second quarter. _____ √

   ii. From the computer output, $s\{Y_{h(new)}\}$=5.79 when X_{h1}=54.
       Construct a 95 percent prediction interval for sales in the
       second quarter of 1987. (Note: t(.975; 46) = 2.013).

   iii. Interpret this prediction interval. _____

                                                                √

D. REGRESSION MODELS WITH AUTOCORRELATED ERROR TERMS
(Text: Chapter 25, Section 25.5)

Review of Basic Concepts

D.1 Explain what is meant by the phrase "autocorrelated error terms."

_____

D.2 The first-order autoregressive error model assumes that the error
terms ϵ_t are composed of two parts. The component $\rho\epsilon_{t-1}$, where
$\rho > 0$, indicates that ϵ_t contains part of the error term from
period _____. The component u_t is called a(n) _____ _____
term. The u_t's are _____ of one another and are _____
distributed with mean _____ and _____ variance.

D.3 In the first-order autoregressive error model, the parameter ρ is
called the _____ parameter.

D.4 Does it appear as if ρ is positive or negative in the following
annual time series: 100, 15, 110, 10, 127, 18, 142, 31? _____

D.5 The regression model with independent error terms can be thought of
as a special case of the first-order autoregressive error model
with ρ = _____.

D.6 In the Durbin-Watson test for positive autocorrelatlon, the
appropriate alternatives are H_0: _____ and H_1: _____.
Alternative H_0 is selected if the value of the test statistic,
which is denoted by _____ is _____ and
alternative H_1 is selected if it is _____.

D.7 An annual time series with η - 20 observations and one independent
variable is fit to trend model (25.10) in your Text. It is desired
to test whether or not the error terms in this model are positively

correlated, with α = .05. Explain in words how the test statistic for this Durbin-Watson test is computed. _____

_____

D.8 For the situation in D.7, the appropriate decision rule is

_____ .

D.9 For the situation in D.7 and D.8, what conclusion would be reached if the value of the test statistic had been d = 1.647? _____

_____

D.10 For the situation in D.7 and D.8, what conclusion would be reached if the value of the test statistic had been d = 1.332? _____

_____

D.11 In the method of first differences, X_t' = _____ and Y_t' = _____ .

D.12 The method of first differences with regression through the origin usually is used when the error terms are _____ correlated and ρ is near _____ .

D.13 In a regression model with regression through the origin, the intercept parameter β_0 = _____ .

D.14 The estimated first differences model in a study is $\hat{Y}_t'$ = $1.2_{X_t'}$. A forecast of X_t' for t = 35 is 620. Thus, Y'_{35} = _____ . Interpret this result. _____

_____

»Problem 6

The following regression model was used by an auto industry analyst to predict quarterly deseasonalized automobile sales in one geographic region for the period 1975 to 1984, where t=1 at Quarter 1 of 1976.

$$\hat{Y}_t = \beta_0 + \beta_1 X_t + \epsilon_t$$

where
Y_t = deseasonalized automobile sales ($ millions) in the region in quarter t.
X_t = deseasonalized total personal disposable income ($ millions) in the region in quarter t.

a. Using the simple linear regression model (18.3) in your Text, the analyst obtained the estimated regression function $\hat{Y}_t =$ 10.997 + .0495X_t. This function was deemed by the administrator to provide a good fit (e.g., $r^2 = .856$), but yielded the residuals in Table 25.5 (which also shows the observations on Y_t and X_t).

Table 25.5
OBSERVATIONS AND RESIDUALS FOR STUDY OF
DESEASONALIZED QUARTERLY AUTOMOBILE SALES
(See Problem 6)

t	Y_t	X_t	e_t	t	Y_t	X_t	e_t	t	Y_t	X_t	e_t
1	50.0	1035.2	-12.2378	14	99.0	1451.3	16.1660	27	107.9	2078.6	-5.9842
2	51.5	1105.2	-14.2027	15	98.1	1496.2	13.0435	28	95.9	2109.8	-19.5286
3	58.9	1109.4	- 7.0106	16	99.3	1542.7	11.9419	29	106.2	2132.0	-10.3274
4	62.8	1134.5	- 4.3530	17	99.5	1587.5	9.9243	30	105.1	2156.8	-12.6550
5	69.8	1163.7	1.2017	18	93.5	1624.0	2.1177	31	118.1	2195.8	-11.5854
6	71.4	1180.8	1.9553	19	98.4	1674.3	4.5279	32	115.3	2237.5	-6.4495
7	73.2	1203.3	2.6415	20	94.9	1714.9	-0.9817	33	115.3	2261.4	-7.6325
8	75.8	1229.6	3.9398	21	97.9	1771.7	-0.7932	34	128.4	2302.9	3.4133
9	83.4	1255.2	10.2726	22	80.6	1789.8	-18.9892	35	132.0	2367.4	3.8207
10	83.8	1291.9	8.8560	23	90.1	1846.0	-12.2710	36	141.7	2428.6	10.4914
11	85.0	1335.5	7.8979	24	94.3	1908.0	-11.1398	37	147.7	2502.2	12.8483
12	87.0	1373.5	8.0170	25	105.4	1972.5	-3.2325	38	152.3	2554.3	14.8695
13	86.4	1405.7	5.8231	26	98.5	2006.	-11.7907	39	148.6	2606.4	8.5906
								40	150.7	2644.5	8.80478

i. Based on visual inspection of these residuals, does it appear that the error terms in this model satisfy the assumption of independence over time? _____ Explain. _____

_____

ii. The analyst noted on the computer output for this regression analysis that the Durbin-Watson test statistic was d = .369. Conduct a test to determine whether or not the error terms in this model are positively correlated, using $\alpha = .01$. Specify the alternatives, determine the decision rule and state the resulting decision. √

b. Based on the test results in part (a.ii.), the analyst decided to use the method of first differences for these data.

   i. Obtain the differences for 1975.

t	Y_t'	X_t'
1	_____	_____
2	_____	_____
3	_____	_____
4	_____	_____

   ii. The following calculational results are obtained from the full set of nd = 39 first differences (computations not shown): $\Sigma X_t'Y_t'$ = 5378.15 and $\Sigma(X_t')^2$ = 76997.35. State the fitted first differences model. _____.

   iii. The Durbin-Watson test statistic for the model was d = 2.666. What conclusion can be reached about the autocorrelation of error terms in this model, using α = .01? _____ Explain. ✓
   _____

   iv. A government report has just forecast that deseasonalized total personal disposable income in this region will increase by \$31.6 million between quarters 40 and 41. Obtain a point estimate forecast of deseasonalized automobile sales in quarter 41. _____ ✓

Answers to Chapter Reviews and Problems

A.1 irregular or random

A.2 Y_t; A_t

A.3 A_6 = .25(510) + .75(550) = 540

A.4 F_7; F_8; F_7 = 540; F_8 = 540

A.5 If several observations form the immediate past are available, A_0 may be set equal to the mean of those observations. If not, A_0 can be specified using managerial judgement.

A.6 smoothing constant; increasing

A.7 decreased

A.8 A_3 = .2(615) + .8(601.20) = 603.96

A.9 t; all

A.10 A_3 = .2(615) + .2(.8)(590) + .2($.8^2$)(600) + .2($.8^3$)(600) = 603.96

A.11 weighted average; .2; .2(.8) = .16; .2($.8^2$) = .128; ($.8^3$) = .512

A.12 .10; .10($.90^4$) = .0656; .10($.90^9$) = .0387

Problem 1

a. i. A_{115} = .2(2415) + .8(2440) = 2435; thus, F_{116} = 2435.

ii. The smoothed estimate for July, A_{115} = 2435 must be carried over. This smoothed estimate, together with August quantity demanded, Y_{116} = 2420, and the smoothing constant, are used to prepare the September forecast.

iii. A_{116} = .2(2420) + .8(2435) = 2432; Thus, F_{117} = 2432

b. i. With a = .3, F_{116} = A_{115} = .4(2415) + .7(2440) = 2432.5, and F_{117} = A_{116} = .3(2420) + .7(2432.5) = 2428.75

B.1 A_t; B_t; mean level

B.2 trend adjustment

B.3 A_{20} = .10(4538.40) + .90(4521.80 + 14.30) = 4536.33

B.4 B_{20} = .25(4536.33 - 4521.80) + .75(14.30) = 14.3633

B.5 F_{21} = 4536.33 + 1(14.36) = 4550.69;
F_{22} = 4536.33 + 2(14.36) = 4565.05;
F_{23} = 4536.33 + 3(14.36) = 4579.41

B.6 smoothed estimate; trend estimate; seasonal index estimates

B.7 c; large

B.8 A_{24} = .05(858.8/1.357) + .95(643.5 - 8.70) = 634.7

B.9 B_{24} = .20(634.7 - 643.5) + .80(-8.7) = -8.72;
C_{24} = .45(858-8/634.7) + .55(1.357) = 1.355

B.10 $F_{25} = [634.7 + 1(-8.72)](.845) = 529.0;$
$F_{26} = [634.7 + 2(-8.72)](.812) = 501.2;$
$F_{27} = [634.7 + 3(-8.72)](.982) = 597.6;$
$F_{28} = [634.7 + 4(-8.72)](1.355) = 812.8$

Problem 2

a. $A_{14} = .15(46.4) + .85(46.002 + 1.2193) = 47.098;$
$S_{14} = .25(47.098 - 46.002) + .75(1.2193) = 1.1885;$
The remaining estimates are similarly obtained, yielding:

t	Y_t	A_t	B_t
14	46.4	47.098	1.1885
15	48.9	48.379	1.2116
16	49.5	49.577	1.2082

b.i.

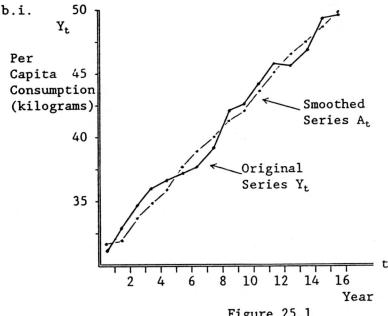

Figure 25.1

ii. Yes; the smoothed estimates track the time series
observations quite well and, hence, will likely provide
reasonable forecasts for future periods.

c. 1987: $F_{17} = 49.577 + 1(1.2082) = 52.785$
1988: $F_{18} = 49.577 + 2(1.2082) = 51.993$
1989: $F_{19} = 49.577 + 3(1.2082) = 53.202$
1990: $F_{20} = 49.577 + 4(1.2082) = 54.410$

Problem 3

a. A_{19} = .15(65.5/1.3010) + .85(52.760 - .9124) = 51.62 - 2;
   B_{19} = .20(51.622 - 52.760) + .80(-.9124) = -.9575;
   C_{19} = .40(65.5/51.622) + .60(1.3010) = 1.2881
   The remaining estimates are similarly obtained, yielding:

t	Y_t	A_t	B_t	C_t
19	65.5	51.622	-.9575	1.2881
20	57.9	50.305	-1.0294	1.1801
21	26.0	49.410	-1.0025	.5214

b. Fall   Semester: F_{22} = [49.410 + 1(-1.0025)](1.2881) = 62.354
   Spring Semester: F_{23} = [49.410 + 2(-1.0025)](1.1801) = 55.942
   Summer Sessions: F_{24} = [49.410 + 3(-1.0025)](.5214) = 24.194

C.1 independent variables; independent; normally; 0; constant

C.2 Y_t = β_0 + $\beta_1 X_t$ + ϵ_t, where X_t=t and ϵ_t are independent $N(0,\sigma^2)$

C.3 4.3; the linear trend component of sales of the firm is estimated to increase \$4.3 thousand from one quarter to the next.

C.4 -0.7

C.5 coincident; lagged

C.6 185 - 1.734(6.2) = 174; 185 + 1.734(6.2) = 196

C.7 Y_t = β_0 + $\beta_1 X t_1$ + $\beta_2 X t_2$ + $\beta_3 X t_3$ + $\beta_4 X t_4$ + $\beta_5 X_{t-1,5}$ + $\beta_6 X_{t-1,6}$ + $\beta_7 X_{t-2,6}$ + ϵ_t
   where X_{t1} = t
   X_{t2}, X_{t3}, X_{t4} are indicator variables for quarterly seasonal effects
   $X_{t-1,5}$ is building permits issued lagged one quarter
   $X_{t-1,6}$, and $X_{t-2,6}$ are disposable earnings lagged one and two quarters, respectively
   ϵ_t's are independent $N(0,\sigma^2)$

C.8 distributed time lag

Problem 4

a. i.

Year	X_t	Y_t	$\hat{Y}_t$	$Y_t - 2\hat{Y}_t$
1983	7	3.57	3.608	-.038
1984	8	3.80	3.789	.011
1985	9	3.90	3.970	-.070
1986	10	4.23	4.151	.079

ii.

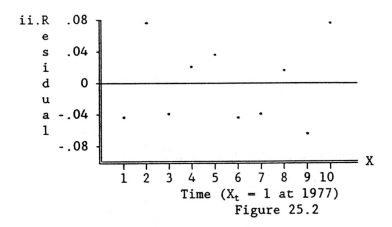

Figure 25.2

iii. Yes; the residual plot over time suggests that the error terms are not related to one another. Thus, the assumption of independent error terms appears reasonable here.

b. i. $b_1 = .181$; $s(b_1) = .0063$; $t(.95;8) = 1.860$; the confidence limits are $.181 \pm 1.860(.0063)$; thus, the 90 percent confidence interval for β_1 is $.169 \le \beta_1 \le .193$.

ii. With 90 percent confidence, we can state that expected ridership increases between .169 million and .193 million from one year to the next.

c. $Y_{11} = 2.341 + .181(11) = 4.332$; $s(Y_{h(new)}) = .070$; $t(.95;8) = 1.860$; the prediction limits are $4.332 \pm 1.860(.070)$; thus, the 90 percent prediction interval for ridership in 1987 is: $4.202 \le Y_{h(new)} \le 4.462$.

Problem 5

a. Alternatives: H_0: $\beta_1 = \beta_2 = \beta_3 = \beta_4 = \beta_5 = 0$
                    H_1: not all $\beta_k = 0$

         F*:   MSR = 442.70;
                    MSE = 13.57;
                    Thus, F* = 442.70/13.57 = 32.62

Decision Rule: F(.99; 5, 46) ≐ 3.45; thus, the decision rule
                    is:
                    If F* ≤ 3.45, conclude H_0;
                    If F* > 3.45, conclude H_1

    Decision: Since F* = 32.62 > 3.45, conclude H_1, there is
                    a regression relation between quarterly
                    shipments of lumber in the region and the
                    independent variables in this model.

b. $\hat{Y}_t = 11.712 + .383X_{t1} + 6.77X_{t2} + 3.245X_{t3} - .423X_{t4} + .825_{Xt-4,5}$

c. i. $\hat{Y}_{54} = 11.712 + .383(54) + 6.77(1) + 3.245(0) - .423(0) +$
      $.825(31.6) = 65.234$

  ii. $\hat{Y}_{54} = 65.234$; $s\{Y_{h(new)}\} = 5.79$; $t(.975; 46) = 2.013$; the
      prediction limits are 65.234 ± 2.013(5.79); thus, the 95
      percent prediction interval for shipments of lumber in the
      region in the second quarter of 1987 is:
      $53.58 \leq Y_{h(new)} \leq 76.89$.

 iii. With 95 percent confidence, we can state that shipments of
      lumber in the region in the second quarter of 1987 will be
      between $53.58 million and $76.89 million.

D.1 When the error terms corresponding to different periods in a time
series regression model are correlated with one another, they are
said to be autocorrelated error terms.

D.2 1 - 1; random disturbance; independent; normally; 0 ; constant

D.3 autocorrelation

D.4 negative, since consecutive observations alternate systematically between large and small values

D.5 $\rho = 0$

D.6 H_0: $\rho \le 0$; H_1: $\rho > 0$; d; large; small

D.7 The difference between each of the nineteen pairs of adjacent residuals (i.e., the difference $e_t - e_{t-1}$) is computed. These nineteen differences are squared, and then summed. Finally, that sum is divided by the sum of the twenty squared residuals, yielding the test statistic, d.

D.8 If $d > 1.41$, conclude H_0; if $d < 1.20$, conclude H_1; if $1.20 \le d \le 1.41$, the test is inconclusive.

D.9 Since $d = 1.647 > 1.41$, conclude $H_0(\rho \le 0)$, the error terms are not positively correlated.

D.10 Since $1.20 < d = 1.332 < 1.41$, the Durbin-Watson test is inconclusive. However, given this uncertainty it would be prudent to conclude that the error terms may be positively correlated, and to act accordingly.

D.11 $X_t' = X_t - X_{t-1}$; $Y_t' = Y_t - Y_t-1$

D.12 positively; 1.0

D.13 $\beta_0 = 0$

D.14 $\hat{Y}'_{35} = 1.2(620) = 744$; the expected value of the dependent variable is estimated to increase 744 between periods 34 and 35.

Problem 6

a. i. No; there appears to be a succession of negative residuals (periods 1 to 4), then positive residuals (periods 5 to 19), then negative residuals (periods 20 to 33), and finally positive residuals (periods 34 to 40).

ii. Alternatives: H_0: $\rho \leq 0$ (error terms are not positively
   correlated)
   H_1: $\rho > 0$ (error terms are positively
   correlated)

   d: d = .369

Decision Rule: α = .01; n = 20; (p - 1) = 1; from Table C-12,
   d_L = 1.25 and d_U = 1.34; thus, the decision
   rule is:
   If d > 1.34, conclude H_0; if d < 1.25, conclude
   H_1
   If 1.25 ≤ d ≤ 1.34, the test is inconclusive

Decision: Since d = .369 < 1.25, conclude $H_0(\rho > 0)$, the
   error terms in this model are positively
   correlated.

b. i.

i	Y_t'	X_t'
1	---	---
2	1.5	70.0
3	7.4	4.2
4	3.9	25.1

ii. b1 = 5378.15/76997.35 = .0698; thus, the fitted first
   differences model is: $\hat{Y}_t'$ = .0698Xt'.

iii. With α = .01, n = 39, and (p - 1) = 1, we obtain, from Table
   C-12, d_L = 1.24 and d_U = 1.34. Since d = 2.666 > 1.34,
   conclude $H_0(\rho \leq 0)$, the error terms in the first differences
   model are not positively correlated.

iv. $\hat{Y}'_{41}$ - .0698(31.6) = 2.206 is the estimated change in
   deseasonalized automobile sales between quarters 40 and 41.
   Thus, since Y_{40} = 150.7 (from Table 25.5), the forecast for
   quarter 41 is 150.7+2.206 = 152.9, or $152.9 million.

26

PRICE AND QUANTITY INDEXES

A. PRICE RELATIVES AND LINK RELATIVES
(Text: Chapter 26, Section 26.1)

Review of Basic Concepts

A.1 A manager is studying changes in the price per cubic foot of natural gas, which is used to heat a plant. If the base year is 1972, the price relative for 1986 equals the _____ price divided by the _____ price, expressed as a percent. The link relative for 1986 equals the _____ price divided by the _____ price, expressed as a percent.

A.2 The price relative (1972=100) for a commodity in two consecutive years are 200 and 250. During this time span, the percent points of change in the price relatives was _____ and the percent change was _____.

A.3 For the situation in A.2, which of the two computations measures the relative change in the price relatives between the two years?

_____

A.4 The price relative (1977=100) for a cleaning fluid in 1986 is 82. Does this mean that the price of the item decreased from 1985 to 1986? _____ Explain. _____

_____

A.5 The link relatives for the prices of an item in 1984, 1985, and 1986 are 125, 92, and 144, respectively. Therefore, the price of the item _____ between 1984 and 1985, and it _____ between 1985 and 1986.

A.6 The prices of a certain product in two states in 1986 and in the base year were as follows:

	Minnesota	Wisconsin
Base Year	$132	$100
1986	$330	$275

Thus, the 1986 price relative for Minnesota is _____ and the 1986 price relative for Wisconsin is _____ .

A.7 For the situation in A.6, the price of the product in 1986 was larger in the state of _____ and the 1986 price relative was larger in the state of _____ . √

»**Problem 1**

List prices for an indoor play gym during the period 1981-1986 appear in the following table. A buyer for the toy department of a large mail-order catalog firm is interested in analyzing the price series for this item.

Year	List Price	Price Relatives 1981 = 100	1983 = 100	Link Relatives
1981	26.50	100.0	_____	
1982	28.00	105.7	_____	105.7
1983	30.00	113.2	_____	107.1
1984	29.50	_____	_____	_____
1985	34.00	_____	_____	_____
1986	36.50	_____	_____	_____

a. Determine the price relatives for 1984 to 1986, with 1981 as the base period. √

b. Determine the price relatives for 1981 to 1986, with 1983 as the base period. √

c. i. What were the percent points of change and percent change in the price relatives between 1984 and 1985, with 1981 as the base period? _____

   ii. What were the percent points of change and percent change in the price relatives between 1984 and 1985, with 1983 as the base period? _____ √

iii. What generalization is suggested by your results in (i.) and
(ii.)? _____
_____ √

d. i. Determine the link relatives for this price series.

ii. Explain the meaning of the link relative for 1984. _____
_____
_____

iii. What is the relationship between the link relative for 1985
and the percent change in the price relatives between 1984
and 1985? _____
_____ √

B. PRICE INDEXES BY METHOD OF WEIGHTED AGGREGATES
(Text: Chapter 26, Section 26.2)

Review of Basic Concepts

B.1 A price index is a summary measure of _____ _____
changes over time in a set of items.

B.2 The group or list of items included in a price index is called the
_____ of items.

B.3 In developing a local index of food prices, an analyst concluded it
was economically infeasible to include all food items. Instead, a
sample of 15 food items is to be used. Would you recommend that
the analyst select a simple random sample of 15 items from all the
food items available in the local area? _____ Explain.
_____

B.4 When using the method of weighted aggregates, how is the importance
of each item usually measured? _____
_____

B.5 In constructing a price index, $P_{2,3}$ refers to the price of the item
in the schedule in the _____ period. The price of the same
item in the base period is denoted by _____ .

B.6 The symbol Q_{2a} refers to the quantity of the _____
item in the schedule that is consumed in a(n) _____ period.

B.7 A three-item price index is to be constructed using the following
data, with 1985 as the base year:

Item	Quantity	Unit Prices 1985	1986
1	300	6.00	6.50
2	450	25.00	18.00
3	250	9.50	9.50

Obtain each of the following:

a. Q_{1a} = _____ ; Q_{2a} = _____ ; Q_{3a} = _____

b. $P_{1,0}$ = _____ ; $P_{2,0}$ = _____ ; $P_{3,0}$ = _____

c. $P_{1,1}$ = _____ ; $P_{2,1}$ = _____ ; $P_{3,1}$ = _____

d. $\Sigma P_{i0}Q_{ia}$ = _____ ; $\Sigma P_{i0}Q_{ia}$ = _____

e. I_{85} = _____ ; I_{86} = _____

B.8 Refer to Table 26.2 in your Text. The percent change in the price
index between 1984 and 1985 is 7.3 percent with lg83 as the base
year. Suppose the base period had been 1985 rather than 1983. If
so, then I_{83} = _____ , I_{84} = _____ ,
I_{85} = _____ and the percent change in the price index
between 1984 and 1985 is computed as _____ . Thus,
percent changes in a price index are not affected by the choice of
the _____ _____ .

B.9 A Laspeyres price index uses as quantity weights the quantities
consumed in the _____ period while a Paasche price index
uses the quantities consumed in each given period t.

B.10 Identify the disadvantage(s) of frequent changes in the schedule of
items and/or quantity weights in a price index. _____
_____

B.11 Identity the disadvantage(s) of frequent changes in the schedule of
items and/or quantity weights in a price index. _____
_____ √

»Problem 2

The operator of a regional grain elevator is interested in developing a price index for the period 1983 - 1986 for the four major grains handled. Relevant information is presented in the following table, where the quantity weights are measured as millions of bushels handled in 1983 and the unit prices are measured as the price per bushel each year.

				Unit Prices		.
	Schedule of Quantity		1983	1984	1985	1986
i	Items	Q_{ia}	P_{i0}	P_{i1}	P_{i2}	P_{i3}
1	Corn	3.40	2.28	2.16	2.40	2.12
2	Wheat	1.20	2.74	2.66	3.10	2.86
3	Oats	1.20	1.28	1.48	1.28	1.36
4	Barley	.40	1.86	2.00	2.44	2.03

a. Compute the following table of computations needed to obtain the price index by the method of weighted aggregates. √

i	$P_{i0}Q_{ia}$	$P_{i1}Q_{ia}$	$P_{i2}Q_{ia}$	$P_{i3}Q_{ia}$
1	7.752	7.344	8.160	7.208
2	3.288	3.192	3.720	3.432
3	1.536	____	____	____
4	.744	____	____	____
Total	13.320	____	____	____

b. Obtain the price index series with 1983 as the base year.

I_{83} = _____ ; I_{84} = _____ ;

I_{85} = _____ ; I_{86} = _____ √

c. Interpret the value of I_{86} computed in (b.). _____
_____

d. By what percent did grain prices at this elevator change between 1985 and 1986? _____ √

C. PRICE INDEXES BY METHOD OF WEIGHTED AVERAGE OF RELATIVES
(Text: Chapter 26, Section 26.3)

Review of Basic Concepts

C.1 In the method of weighted aggregates, the importance of each item in the schedule is measured in units of the quantity typically consumed. In the method of weighted average of relatives, the importance of each item is usually measured in units of
_____ _____. Thus, weights used in the latter method are called _____ weights. ✓

C.2 The symbol V_{4a} is used to denote the _____ _____ for the _____ item in the schedule. If the typical quantity consumed for this item is 1,500 and the typical price is \$3.00, V_{4a} = _____.

C.3 With the method of weighted average of relatives, the price index for any period is a weighted average of the _____ _____ using _____ _____ as the appropriate weights.

C.4 In a two-item price index, V_{1a} = \$300 and V_{2a} = \$350. If the price relatives for the two items in period 4 are 92 and 105, respectively, the price index in period 4 by the method of weighted average of relatives is _____.

C.5 The method weighted aggregates and the method of weighted average of relatives yield identical results if, in the latter method, the value weights V_{ia} are obtained by multiplying typical quantities consumed times _____ _____ _____ ✓

»**Problem 3**

The office manager for a government agency is developing a price index for stationery products used in the agency. The two items used in the index are #8 blank envelopes and 81 X 11 inch letterhead. The manager has estimated that the agency uses 24 boxes of envelopes and 60 reams of letterhead in a typical year. The following table shows the unit price of these two items for the period 1982 - 1986.

i	Schedule of Items	1982	1983	1984	1985	1986
				Unit Price		
1	#8 Blank Envelopes	4.50	4.65	4.90	5.00	5.20
2	8½x11" Letterhead	8.30	8.35	8.50	8.55	8.80

a. i. Compute the price relatives for each of the two items during the period 1982 - 1986, with 1982 as the base year. √

   ii. Which of the two items in the schedule experienced the greater relative price increase between 1982 and 1986?

   _____ √

b. The mean unit prices of the envelopes and letterhead during the period 1982 - 1986 were $4.85 and $8.50, respectively. Using these mean prices and the typical quantities consumed per year, obtain the weights for the two items.

   V_{1a} = _____ ; V_{2a} = _____ √

c. Using the price relatives from part (a.i) and the value weights from part (b.), obtain the price index series for the period 1982 - 1986 by the method of weighted average of relatives, with 1982 as the base year. √

d. By what percent did stationery prices increase between 1982 and 1986? _____

e. Is the price index series in (c.) the same as that which would have been obtained by the method of weighted aggregates? _____ Explain. _____
   _____ √

D. OTHER ISSUES IN THE CONSTRUCTION AND USE OF PRICE INDEXES
(Text: Chapter 26, Sections 26.4 to 26.6)

Review of Basic Concepts

D.1 It has been proposed to use a year in which major crop failures have driven coffee prices to record high levels as the base year for a price index for coffee. Why would this selection of a base year in inappropriate? _____

D.2 In the period 1976 - 1978 is to be used as the base period for a particular price index, the index numbers should be calculated in such a way that the indexes for 1976, 1977, and 1978 have an average value of _____.

D.3 Suppose a price index series appears as follows:

Year:	1983	1984	1985	1986
Price Index:	100	108	112	120

The base year for this index is 1983. To shift the base year to 1986, each index number is divided by _____. Thus, with 1986 as the base period, the index number for 1984 is _____.

D.4 If the appropriate data are available, the base period for a price index can be shifted by recalculating the index. If not, only the short-cut method in D.3 can be used. The results of these two procedures are identical if the index was originally compiled using the method of _____ _____ with fixed _____ weights.

D.5 Index numbers for a particular price index were 100 in 1972 and 160 in 1978. The index was revised in 1978, yielding index numbers of 100 in 1978 and 145 in 1986. To splice these two series into one with 1978 as the base period, each index number in the original series is divided by _____. Thus, the index number for 1972 in the spliced series is _____.

D.6 The specification for one item in a food price index calls for "one package of fresh grapefruit." Why is this an inadequate specification for the item? _____

_____

Develop an improved specification for this item. _____

_____

D.7 Identify which, if any, of the following quality changes appear to be major ones affecting the product and, hence, ones that should be reflected in a price index for automobile supplies.

a. A change in the formulation of engine oil which allows it to maintain fluidity at temperatures two degrees colder than previously attainable. _____

b. A change in the formulation of engine oil which allows it to be replaced just once every 12,000 miles rather than once every 3,000 miles. _____

c. A change in the material used to manufacture windshield wiper blades which reduces the expected life of the blades by three percent. _____

D.8 Based on a recent survey of consumer spending patterns, it was found that purchases of hamburger and chicken comprise 27.3 percent of all meat, poultry and fish purchases and 5.4 percent of all food purchases. Further, purchases of meat, poultry and fish comprise 19.1 percent of all food purchases. If hamburger and chicken are the only meat, poultry and fish related items in a schedule of items for a food price index, what should be their combined weight in the schedule as a percent of the total weight for all items? _____ Explain. _____

_____

D.9 Suppose a person who began working full-time in 1967 lists his annual income for each year between 1967 and 1986. These annual income figures are expressed in _____ dollars. If this person divides each year's income by the Consumer Price Index (1967 = 100), the annual income figures are now expressed in _____ dollars or _____ dollars, and represent _____ earnings for that person over the period 1967 - 1986.

D.10 For the situation in D.9, suppose the person converts his real earnings to percent relatives with 1967 as the base year and he finds that the relative for 1986 is 92.7. Thus, his real earnings in 1986 have _____ to _____ percent of what they were in _____.

D.11 The manager of a shopping center is studying growth in retail sales of the center over time. She has data on total annual sales volume (in dollars)-since the center was opened. Why might she wish to use a price index to adjust this time series? _____

_____

_____

D.12 A secretarial union has a contract which adjusts hourly wage rates proportionately to increases in the Consumer Price Index. This _____ of _____ adjustment is an example of a(n) _____ _____ in a union contract. √

»Problem 4

The owner of a service station has obtained a regional price index for gasoline for the period 1972 - 1985 with 1967 as the base year. This price index series appears in Table 26.1a.

Table 26.1
GASOLINE PRICE INDEX AND SALES VOLUME FOR
STUDY OF SERVICE STATION SALES
(See Problem 4)
(a)
Gasoline Price Index

Year	Price Index (1967=100)	Price Index (1977=100)	Year	Price Index (1967=100)	Price Index (1977=100)
1972	107.6	_____	1979	265.6	_____
1973	118.1	_____	1980	369.1	_____
1974	159.9	_____	1981	410.9	_____
1975	170.8	_____	1982	389.3	_____
1976	177.9	_____	1983	376.3	_____
1977	188.2	_____	1984	370.2	_____
1978	196.3	_____	1985	373.3	_____

(b)
Annual Gasoline Sales ($ thousands) for Period 1972 - 1980

Year	Annual Sales	Annual Price Index (1977=100)	Percent Sales at 1977 Prices	Percent Relative (1977=100)	Link Relative
1977	1,752.0	_____	_____	_____	_____
1978	1,950.2	_____	_____	_____	_____
1979	2,549.0	_____	_____	_____	_____
1980	3,517.6	_____	_____	_____	_____
1981	3,881.6	_____	_____	_____	_____
1982	3,572.3	_____	_____	_____	_____
1983	3,388.9	_____	_____	_____	_____
1984	3,305.3	_____	_____	_____	_____
1985	3,395.0	_____	_____	_____	_____

a. i. Calculate the percent change in regional gasoline prices from the preceding year for each of the following selected years.

1974 _____ ; 1979 _____ ;

1980 _____ ; 1983 _____

ii. By what percent did regional gasoline prices increase between 1967 and 1985? _____ √

b. Shift the base of this price index series to 1977. √

c. Total annual gasoline sales (in dollars) for the service station for the period 1977 - 1985 appear in Table 26.1b. Using the price index for gasoline with 1977 as the base year, express total annual gasoline sales for the period 1977 - 1985 in 1977 prices. Also, determine the percent relatives (1977 = 100) and link relatives for total annual gasoline sales at 1977 prices. √

d. Using the results from (c.), discuss the year-to-year relative changes in quantities of gasoline sold by the service station during the period 1977 - 1985. _____

_____

_____ √
_____

E. QUANTITY INDEXES AND MAJOR PUBLISHED INDEXES

(Text: Chapter 26, Sections 26.7 and 26.8)

»Problem 5

A firm sells three models of an industrial generator. The production manager for the firm wishes to construct a quantity index for the production of generators during the period 1984 - 1986 using the method of weighted aggregates. The 1985 unit values added (in $ thousands), denoted by P_{ia}, will be used as weights, with 1984 as the base year for the index. These weights, together with the quantity of each model produced in 1984, 1985 and 1986, appear in the following table.

			Quantities Produced		
	Schedule	Weights	1984	1985	1986
i	of Items	P_{ia}	Q_{i0}	Q_{i1}	Q_{i2}
1	Model A	8.0	475	422	422
2	Model B	12.5	320	355	390
3	Model C	20.0	114	214	276

a. Complete the following computations, which are required to construct the quantity index. ✓

i	$Q_{i0}P_{ia}$	$Q_{i1}P_{ia}$	$Q_{i2}P_{ia}$
1	3,800	3,376	3,376
2	4,000		
3	2,280		
Total	10,080		

b. Compute the quantity index series for the period 1984 - 1986 with 1984 as the base year.

c. Interpret the index number for 1986. _____

_____ ✓

Answers to Chapter Reviews and Problems

A.1 1986; 1972; 1986; 1985

A.2 50; 100(50/200) = 25

A.3 percent change (25 percent)

A.4 No; it means the price of the item in 1986 was less than the price of the item in 1977, the base year. One cannot tell from the information given whether the price increased or decreased from 1985.

A.5 decreased; increased

A.6 100(330/132) = 250; 100(275/100) = 275

A.7 Minnesota, Wisconsin

Problem 1

a. and b.

Year	Price	List Price Relatives 1981=100	1983=100
1981	26.50	100.0	88.3
1982	28.00	105.7	93.3
1983	30.00	113.2	100.0
1984	29.50	111.3	98.3
1985	34.00	128.3	113.3
1986	36.50	137.7	121.7

c. i. Percent points of change was 128.3 - 111.3 = 17.0. Thus, percent change was 100(17/111.3) = 15.3.

ii. Percent points of change was 113.3 - 98.3 = 15.0. Thus, percent change was 100(15.0/98.3) = 15.3.

iii. A comparison of the results in parts (i.) and (ii.) suggests that the choice of base period affects the percent points of change in price relatives over time but not the percent change in price relative over time.

d. i.

Year	List Price	Link Relatives
1981	26.50	
1982	28.00	105.7
1983	30.00	107.1
1984	29.50	98.3
1985	34.00	115.3
1986	36.50	107.4

ii. The price of the play gym in 1984 was 98.3 percent as great as the price in 1983.

iii. Both measure the relative change in the price of the play gym between 1984 and 1985. Thus, the percent change in the price relatives (15.3 percent) and the link relative (115.3) both indicate a price increase of 15.3 percent between 1984 and 1985.

B.1 relative price

B.2 schedule

B.3 No. If simple random sampling is used, it is possible that the schedule will contain too many of one type of food item (e.g., canned goods) and too few of another (e.g., produce). A sampling plan should be used which insures that the schedule will be representative of all food items available.

B.4 Importance is usually measured in terms of the typical quantity of each item that is consumed per period.

B.5 second; third; $P_{2,0}$

B.6 second; typical or average

B.7 a. $Q_{1a} = 300$; $Q_{2a} = 450$; $Q_{3a} = 250$

b. $P_{1,0} = 6.00$; $P_{2,0} = 25.00$; $P_{3,0} = 9.50$

c. $P_{1,1} = 6.50$; $P_{2,1} = 18.00$; $P_{3,1} = 9.50$

d. $\Sigma P_{i0}Q_{ia} = 6.00(300) + 25.00(450) + 9.50(250) = 15,425$;
$\Sigma P_{i1}Q_{ia} = 6.50(300) + 18.00(450) + 9.50(250) = 12,425$

e. $I_{85} = 100(15,425/15,425) = 100.00$;
$I_{86} = 100(12,425/15,425) = 80.55$

B.8 $I_{83} = 100(16,544.75/19,449.75 = 85.1$;
$I_{84} = 100(18,128.55/19,449.75 = 93.2$;
$I_{85} = 100(19,449-75/19,449.75 = 100.0$;
$100(6.8/93.2) = 7.3$; base period

B.9 base

B.10 The disadvantage of frequent changes are that changes in the price index over time will reflect changes in the composition of the index as well as price changes and that changing the schedule of items and/or quantity weights can be very expensive.

B.11 The disadvantage of infrequent changes is that there is a risk that the price index will become badly outdated.

Problem 2

j	$P_{i0}Q_{ia}$	$P_{i1}Q_{ia}$	$P_{i2}Q_{ia}$	$P_{i3}Q_{ia}$
3	1.536	1.776	1.536	1.632
4	.744	.800	-976	.812
Total	13.320	13.112	14.392	13.084

b. $I_{83} = 100(13.320/13.320) = 100.00$;
   $I_{84} = 100(13.112/13.320) = 98.44$;
   $I_{85} = 100(14.392/13.320) = 108.05$;
   $I_{86} = 100(13.084/13.320) = 98.23$

c. Since $I_{86} = 98.23$, grain prices at this elevator in 1986 were 98.23 percent as great as in 1983.

d. $100[(98.23 - 108.05)/108.05] = -9.09$ percent

C.1 dollar value; value

C.2 value weight; fourth; V4a = 3.00(1,500) = $4,500

C.3 price relatives; value weights or V_{ia}

C.4 $[92(300) + 105(350)]/(300 + 350) = 99.0$

C.5 base period prices

Problem 3

a. i.

	Price Relatives (1982=100)				
i	1982	1983	1984	1985	1986
1	100.0	103.3	108.9	111.1	115.6
2	100.0	100.6	102.4	103.0	106.0

  ii. Based on the price relatives for 1986, the price of envelopes
      in 1986 was 115.6 percent as great as in 1982 while the price
      of letterhead in 1986 was just 106.0 percent as great as in
      1982. Thus, envelopes experienced a greater relative price
      increase between 1982 and 1986.

  b. $V_{1a} = 4.85(24) = \$116.40$; $V_{2a} = 8.50(60) = \$510.00$.

  c.

i	Value Weight V_{ia}	Price Relative Times Value Weight				
		1982	1983	1984	1985	1986
1	116.40	11,640	12,024	12,676	12,932	13,456
2	510.00	51,000	51,306	52,224	52,530	54,060
Total	626.40	62,640	63,330	64,900	65,462	67,516

  Thus
$$I_{82} = 62,640/626.40 = 100.0;$$
$$I_{83} = 63,330/626.40 = 101.1;$$
$$I_{84} = 64,900/626.40 = 103.6;$$
$$I_{85} = 65,462/626.40 = 104.5;$$
$$I_{86} = 67,516/626.40 = 107.8$$

  d. Since $I_{86} = 107.8$, stationery prices increased 7.8 percent
     between 1982 and 1986.

  e. No; the two methods lead to identical results only if the value
     weights in the method of weighted average of relatives are
     computed with base year prices. Since we used mean prices for
     the five-year period and these differed from base year prices,
     the two methods would lead to somewhat different results.

D.1 Users of a price index tend to assume that the chosen base period
represents a period of relatively normal conditions. The proposal
for the coffee price index would select a base year in which supply
and price conditions were clearly abnormal and, hence, is
inappropriate.

D.2 100.0

D.3 1.20; 108/1.20 = 90.0

D.4 weighted aggregates; quantity

D.5 1.6; 100/1.60 = 62.5

D.6 The specification does not contain sufficient detail. It does not, for example, specify the size of the grapefruit, the number in the package or the variety of grapefruit to be priced. One improved specification is, "One dozen, medium size, fresh Florida red grapefruit."

D.7 a. minor; b. major; c. minor

D.8 19.1 percent; since these two items are used to represent all meat, poultry and fish related items, their combined weight in the schedule should reflect the relative importance of all meat, poultry and fish items in a food price index, which is 19.1 percent.

D.9 current; constant; 1967; real

D.10 decreased; 92.7; 1967

D.11 In current dollars, total annual sales volume over time reflects price changes as well as changes in retail sales activity. With an appropriate price index, total annual sales volume can be adjusted so that retail sales activity can be analyzed without the influence of year-to-year price changes.

D.12 cost of living; escalator clause

Problem 4

    a. i. 1974: $100(41.8/118.1) = 35.4$ percent;
          1979: $100(69.3/196.3) = 35.3$ percent;
          1980: $100(103.5/265.6) = 39.0$ percent;
          1983: $100(-13.0/389.3) = -3.3$ percent

      ii. Since $I_{85} = 373.3$, regional gasoline prices increased 273.3 percent between 1967 and 1985.

b.

Year	Price Index (1977=100)	Year	Price Index (1977=100)
1972	57.2	1979	141.1
1973	62.8	1980	196.1
1974	85.0	1981	218.3
1975	90.8	1982	206.9
1976	94.5	1983	199.9
1977	100.0	1984	196.7
1978	104.3	1985	198.4

c.

Year	Annual Sales	Price Index (1977=100)	Annual Sales at 1977 Prices	Percent Relative (1977=100)	Link Relative
1977	1,752.0	100.0	1,752.0	100.0	-
1978	1,950.2	104.3	1,869.8	106.7	106.7
1979	2,549.0	141.1	1,806.5	103.1	96.6
1980	3,517.6	196.1	1,793.8	102.4	99.3
1981	3,881.6	218.3	1,778.1	101.5	99.1
1982	3,572.3	206.9	1,726.6	98.6	97.1
1983	3,388.9	199.9	1,695.3	96.8	98.2
1984	3,305.3	196.7	1,680.4	95.9	99.1
1985	3,395.0	198.4	1,711.2	97.7	101.9

d. By examining annual sales at 1977 prices, we see that quantity
sold increased in 1978 but decreased in each succeeding year
until 1985, during which it once again increased. The link
relatives show decreases between about 1.0 percent to 3.5
percent annually during the period 1979 - 1984, and an increase
of 1.9 percent in 1985. Based on the percent relative for 1985,
quantity sold in 1985 was just 97.7 percent of that sold in
1977.

Problem 5

a.

i	$Q_{i0}P_{ia}$	$Q_{i1}P_{ia}$	$Q_{i2}P_{ia}$
1	3,800	3,376	3,376
2	4,000	4,438	4,875
3	2,280	4,280	5,520
Total	10,080	12,094	13,771

b. $I_{84} = 100(10,080/10,080) = 100.0;$
    $I_{85} = 100(12,094/10,080) = 120.0;$
    $I_{86} = 100(13,771/10,080) = 136.6$

c. The quantity of generators produced by the firm in 1986 was
   136.6 percent as great as in 1984, as measured by value added.

27

BAYESIAN DECISION MAKING WITHOUT SAMPLE INFORMATION

A. DECISION PROBLEMS
(Text: Chapter 27, Section 22.1)

Review of Basic Concepts

A.1 In the discussion of decision making in your Text, controllable variables in decision problems are reflected by the _____ and are denoted by _____. Further, uncontrollable variables are reflected by the _____ and are denoted by _____.

A.2 A retailer must decide whether or not to carry a new fashion item. The profit will depend upon whether the item achieves limited, moderate, or extensive acceptance among local residents. In this decision problem, there are c = _____ acts or strategies. Identify these acts. _____

A.3 For the situation in A.3, there are r = _____ outcome states. Identify these outcome states. _____
_____

A.4 Identify what the symbol C_{ij} denotes in a decision problem. _____
_____

A.5 In a decision problem, probabilities are sometimes assigned to the _____ _____. If so, these probabilities are denoted by _____.

A.6 A tabular presentation of the structural elements of a decision problem is called a(n) _____ _____ while a graphic presentation of the same information is called a(n)
_____ _____.

A.7 There are two types of nodes in a decision tree. If a node is drawn as a square, it is called a(n) _____ node since the branches emanating from it represent _____.
Conversely, if a node is drawn as a circle, it is called a(n) _____ node since the branches emanating from it represent _____ _____. √

»Problem 1

An investor has $10,000 to deposit in a savings and loan bank for a Period of up to six months. The investor is attempting to decide which of Two savings plans to utilize. First, he could deposit the funds in a passbook account which yields 5.75% annual interest, compounded quarterly. If he does, interest earnings over the six months would be $289.57. Alternatively, he could purchase a six-month money market certificate currently available at an annual interest rate of 11.847% with no compounding of interest over the six-month period, in which case interest earnings over the six months would be $592.35. Unfortunately, the decision is complicated by the fact that he may need the $10,000 at the end of three months. If so, he could remove the funds plus $143.75 (three-months) interest from the passbook account but would forfeit all earned interest when cashing the certificate before maturity. Since short-term loan money is extremely tight, the investor has no alternative but to utilize the $10,000 should the need arise.

a. Identify the acts in this decision problem. _____

_____

b. Identify the outcome states in this decision problem. _____

_____ √

c. Construct a payoff table for this investment decision problem, with pay-offs specified in terms of interest earnings. √

d. Construct a decision tree for this investment decision problem. √

B. DECISION MAKING UNDER CERTAINTY AND UNCERTAINTY
(Text: Chapter 27, Sections 27.2 and 27.3)

Review of Basic Concepts

B.1 If the decision maker is either unwilling or unable to assign probabilities to the various outcome states, the problem is one of decision making under _____.

B.2 If there is no uncontrollable variable in a decision problem, the number of outcome states is r = _____ and the problem is one of decision making under _____.

B.3 The following payoff table represents a problem of decision making under _____.

	Act		
Outcome State	A_1	A_2	A_2
S_1	80	70	80
S_2	20	40	30
S_3	-10	20	SO

B.4 For the situation in B.3, act ____ dominates act _____. Hence, act _____ is inadmissible while acts _____ and _____ are admissible.

B.5 "Maximin" and "minimax regret" are two criteria that can be used to select the best act for decision making under _____.

B.6 The maximin criterion seeks to _____ the _____ for each act while the minimax regret criterion seeks to _____ the _____ _____ for each act.

B.7 Consider the following payoff table and associated table of regrets or opportunity losses.

Payoffs (C_{ij})				Regrets (O_{ij})		
	Act				Act	
Outcome State	A_1	A_2		Outcome State	A_1	A_2
S_1	-200	50		S_1	250	0
S_2	500	200		S_2	0	300

In this decision problem, the minimum payoff if act A_1 is selected is _____ while the minimum payoff if act A_2 is selected is _____. Thus, using the maximin criterion, act _____ would be selected.

B.8 For the situation in B.7, explain why $O_{11} = 250$. _____

_____

B.9 For the situation in B.7, the maximum regret if act A_1 is selected is _____ while the maximum regret if act A_2 is selected is _____. Thus, using the minimax regret criterion, act _____ would be selected. ✓

»Problem 2

Recall the investment decision problem in Problem 1.

a. Suppose, for the moment, that the investor was absolutely sure the funds would not be needed until the end of the six month period.

   i. Construct the payoff table for this situation. √

  ii. Under these conditions, is the problem one of decision making under certainty or uncertainty? _____ Explain.

       _____

 iii. Identify the best act if the investor uses the criterion: Maximize the total interest earned over the six-month period.

    _____

b. Return to the original investment decision problem, the payoff table for which is as follows.

Type of Investment

Investment Period	A_1 Passbook Account	A_2 Money Market Certificate
S_1: 3 Months	143.75	0
S_2: 6 Months	289.57	592.35

  i. Determine the best act if the investor uses the maximin criterion. _____ √

 ii. Construct the table of regrets or opportunity losses for this problem and determine the best act if the investor uses the minimax regret criterion. √

C. BAYESIAN DECISION MAKING WITH PRIOR INFORMATION ONLY
(Text: Chapter 27, Section 27.4)

Review of Basic Concepts

C.1 In Bayesian decision making with prior information only, the prior information referred to is expressed in terms of _____ which are assigned to the _____ _____. This type of decision making is also called decision making under _____.

C.2 Explain in words and symbols how the expected payoff for a particular act, $EP(A_i)$, is computed. _____
_____
_____

C.3 Consider the following payoff table.

		Act		
Outcome State	$P(S_i)$	A_1	A_2	A_3
S_1	.4	-10	0	10
S_2	.6	50	25	0

For this situation, $EP(A_1)$ = _____, $EP(A_2)$ = _____ and $EP(A_3)$ = _____.

C.4 In Bayesian decision making with prior information only, the best act is defined as the act which _____ the _____ _____. This best act is called the _____ act and its payoff is denoted by _____.

C.5 For the situation in C.3, the Bayes act is act ____ and BEP = ____.

C.6 Explain what EPPI measures. _____
_____

C.7 For the situation in C.3, EPPI = _____. Why is it that EPPI is not 50, the largest payoff in the table, if EPPI is supposed to measure the result of having perfect information available? _____
_____

C.8 For the situation in C.3, BEP = 26 and EPPI = 34. Therefore,
EVPI = _____ .

C.9 Explain why the expected value of perfect information generally is
not the same as the expected payoff with perfect information.

_____

_____

C.10 The procedure for determining how susceptible the optimal act and
BEP are to variations in the outcome state probabilities or payoffs
is called _____ .

»Problem 3

Refer once again to the investment decision problem introduced in
Problem 1. Suppose the investor believes there is a forty percent chanc
that he will need the funds at three months. Using this assessment, we
obtain the outcome state probabilities shown in the following payoff
table.

		Type of Investment	
Investment Period	A_1 $P(S_i)$	A_2 Passbook Account	Money Market Certificate
S_1: 3 Months	.4	143.75	0
S_2: 6 Months	.6	289.57	592.35

a. i. Determine the expected payoff for each type of investment
   $EP(A_1)$ = _____ : $EP(A_2)$ = _____

   ii. What is the Bayes Act and BEP for this decision problem?
   _____ √

b. Suppose the investor could purchase certain research data on the
basis of which be would know exactly whether or not he would
need the funds at three months.

   i. Determine the expected payoff with this perfect information.
   _____ √

   ii. Compute EVPI for this situation _____ . How is
   this result useful to the investor? _____
   _____ √

c. Suppose the investor suddenly realizes that he may have underestimated the probability that he will need the funds at three months and now believes that it is equally likely that he will need the funds at three months or at six months. Hence, $P(S_1) = .5$ and $P(S_2) = .5$.

    i. Determine the Bayes act and BEP under this probability assessment. _____

   ii. What effect does this variation in the outcome state probabilities (i.e., have on the Bayes act and BEP)? _____
                                                                                    √
_____

D. MULTISTAGE DECISION PROBLEMS
(Text: Chapter 27, Section 27.5)

»Problem 4

A recent college graduate and sports enthusiast inherited a sizable sum of money and is deciding which of two businesses he should acquire. First, he could acquire a sporting goods store (A_1). If the store succeeds the payoff would be $200 thousand. If the store failing the payoff would be -$50 thousand. Second, he could acquire controlling interest in an ailing ski resort (A_2). If the ski resort succeeds the payoff would be $700 thousand, and if it fails the payoff would be -$170 thousand. Probability assessments for success are .6 for the sporting goods store and .3 for the ski resort.

A hotel near the ski resort is also available and could be acquired later if the ski resort turns out to be successful. If the hotel is acquired (B_1) the payoff of the combined ski resort/hotel acquisition would be $800 thousand if the hotel succeeds but only $650 thousand if the hotel fails. The probability that the hotel will succeed given the ski resort is successful is assessed to be .8.

The decision tree for this acquisition problem appears in Figure 27.1.

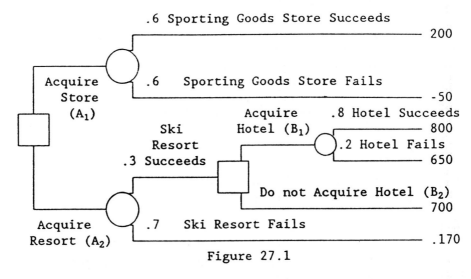

.6 Sporting Goods Store Succeeds
— 200

Acquire
Store
(A₁)   .6   Sporting Goods Store Fails
— -50

Acquire   .8 Hotel Succeeds
Hotel (B₁)   — 800
Ski   .2 Hotel Fails
Resort   — 650
.3 Succeeds

Do not Acquire Hotel (B₂)
— 700

Acquire   .7   Ski Resort Fails
Resort (A₂)   .170

Figure 27.1

a. Using the decision tree in Figure 27.1, determine the Bayes Act and BEP for the hotel acquisition conditioned upon the ski resort being successful. √

b. i. Using the same decision tree, determine the Bayes act and BEP for the entire acquisition decision. √

   ii. Confirm the Bayes act and BEP by forward induction. √

»Problem 5

A corporation has recently acquired a subsidiary which had completed development and testing of a new product. The marketing director of the corporation is in the process of deciding whether to introduce the product regionally (A₁) or drop the product from further consideration (A₂).

If the product is introduced regionally (A₁), there is a .1 probability it will achieve a high level of regional demand. In this case, the marketing director has the option of introducing the product nationally (B₁) or remaining with only regional distribution (B₂). If the former strategy (B₁) is followed, the payoff will be $40.0 million if a high level of national demand is achieved and the payoff will be -$10.0 million if a low level of national demand is achieved. The manager believes that if regional demand is high, there is a .8 probability that national demand will be high. If the latter strategy (B_2) is followed, the payoff will be $6.0 million.

If the product is introduced regionally (A_1) and achieves a low level of regional demand, the probability of which is .9, the payoff will be -$4.0 million. Finally, if the product is dropped from further consideration (A_2), the payoff is $0.

Construct a decision tree for this new product introduction problem and determine the optimal strategy. √

Answers to Chapter Reviews and Problems

A.1 acts or strategies; A_j; outcome states; S_i

A.2 c = 2; carry the item (A_1) and do not carry the item (A_2)

A.3 r = 3; limited acceptance (S_1), moderate acceptance (S_2), and extensive acceptance (S_3)

A.4 C_{ij} denotes the payroll, gain or other consequence realized if act A_j is selected and outcome state S_i prevails.

A.5 outcome states; $P(S_i)$

A.6 payoff table; decision tree

A.7 decision; acts; chance; outcome states

Problem 1

a. The acts are: deposit funds in passbook account (A_1) and purchase money market certificate (A_2).

b. The outcome states are: investment period is three months (S_1) and investment period is six months (S_2).

c.

		Type of Investment	
		---	---
		A_1	A_2
Investment Period		Passbook Account	Money Market Certificate
S_1 : 3 Months		143.75	0
S_2 : 6 Months		289.57	592.35

d.

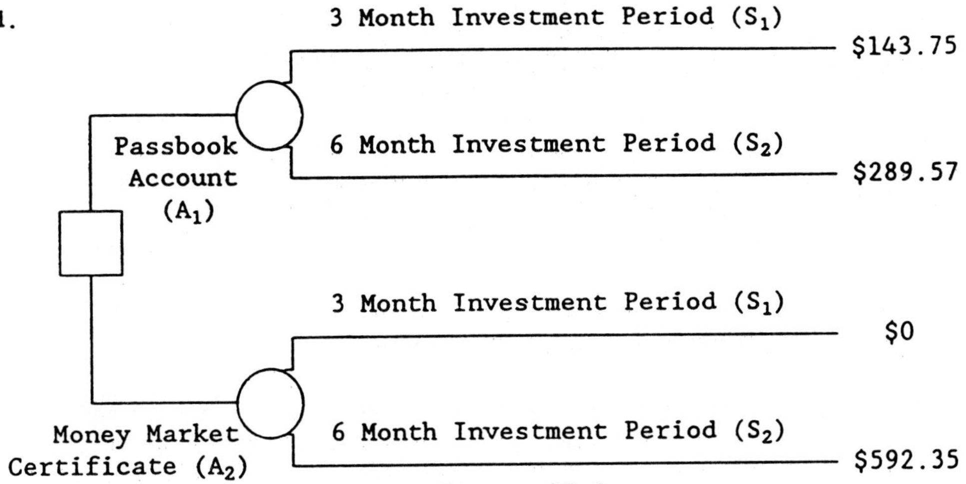

3 Month Investment Period (S_1) — $143.75

Passbook Account (A_1)

6 Month Investment Period (S_2) — $289.57

3 Month Investment Period (S_1) — $0

Money Market Certificate (A_2)

6 Month Investment Period (S_2) — $592.35

Figure 27.2

B.1 uncertainty

B.2 r = 1; certainty

B.3 uncertainty

B.4 A_3; dominates A_1; A_1; A_2 and A_3

B.5 uncertainty

B.6 maximize the minimum payoff; minimize the maximum regret or opportunity loss

B.7 -200; 50; A_2

B.8 The maximum payoff attainable under outcome state S_1 is $C_{12} = 50$. The payoff for act A_1 under outcome state S_1 is $C_{11} = -200$. Thus, the regret or opportunity loss for act A_1 under outcome S_1 is $O_{11} = 50 - (-200) = 250$.

B.9 250; 300; A_1

Problem 2

a. i.

Investment Period	Type of Investment	
	Passbook Account	Money Market Certificate
6 Months	289.57	592.35

   ii. Certainty. In this situation there is only one outcome state since the investment period is certain to be six months.

  iii. The best act is to purchase the money market certificate, with earned interest of $592.35.

b. i. The minimum possible payoffs are $C_{11} = \$143.75$ for act A_1 and $C_{12} = 0$ for act A_2. Since act A_1 maximizes these minimum payoffs, the best act based on the maximin criterion is to deposit the funds in the passbook account.

  ii. The maximum possible payoffs are $C_{11} = \$143.75$ under S_1 and $C_{22} = \$592.35$ under S_2, resulting in the following table of regrets:

Investment Period	Type of Investment	
	A_1 Passbook Account	A_2 Money Market Certificate
S_1: 3 Months	143.75-143.75=0	143.75-0=143.75
S_2: 6 Months	592.35-289.57=302.78	592.35-592.35=0

The maximum possible regrets are $O_{21} = \$302.78$ for act A_1 and $O_{12} = \$143.75$ for act A_2. Since act A_2 minimizes these maximum regrets, the best act based on the minimax regret criterion is to purchase the money market certificate.

C.1 probabilities, outcome states; risk

C.2 The probability assigned to each outcome state is multiplied by the payoff for act A_j under that outcome state. These products are then summed to obtain the expected payoff for act A_j. In symbols, $EP(A_j) = \sum_i C_{ij}P(S_i)$.

C.3 $EP(A_1) = (-10)(.4) + 50(.6) = 26$;
$EP(A_2) = 0(.4) + 25(.6) = 15$;
$EP(A_3) = 10(.4) + 0(.6) = 4$

C.4 maximizes the expected payoff; Bayes, BEP

C.5 A_1; BEP = 26

C.6 EPPI is the expected payoff when the decision maker is able to select the best act using perfect information (i.e., information that will indicate the prevailing outcome state).

C.7 EPPI = $10(.4) + 50(.6) = 34$; the payoff ultimately realized will be either 10 if S_1 Prevails for which the probability is $P(S_1) = .4$) or 50 if S_2 prevails (for which the probability is $P(S_2) = .6$). Perfect information does not allow the decision maker to control which state will prevail, which would be necessary to guarantee a payoff of 50.

C.8 EVPI = 34 - 26 = 8

C.9 BEP is the maximum expected payoff utilizing only prior information in the form of probabilities for the outcome states. EVPI, the expected incremental gain from utilizing perfect information, is the difference between EPPI and BEP. Only if BEP = 0 will EVPI = EPPI - BEP = EPPI - 0 = EPPI.

C.10 sensitivity analysis

Problem 3

a. i. $EP(A_1) = 143.75(.4) + 289.57(.6) = 231.34$;
$EP(A_2) = 0(.4) + 592.35(.6) = 335.41$.

ii. Since act A_2 maximizes the expected payoff, the Bayes act is to purchase the money market certificate and BEP = $355.41.

b. i. EPPI = $143.75(.4) + 592.35(.6) = 412.91$

    ii. EVPI $-$ 412.91 $-$ 355.41 $-$ 57.50; the investor should be willing to pay up to \$57.50 for this perfect information.

c. i. $EP(A_1)$ $-$ 143.75(.5) + 289.57(.5) $-$ 216.66;
$EP(A_2)$ $-$ 0(.5) + 592.35(5) $-$ 296.18;
thus, the Bayes act is act A_2 and BEP $-$ \$296.18.

    ii. The Bayes act has not changed, but BEP has decreased from \$355.41 to \$296.18.

Problem 4

a. $EP(B_1)$ $-$ 800(.8) + 650(.2) $-$ 770;
$EP(B_2)$ $-$ 700;
thus, the Bayes act is act B_1, BEP $-$ \$770 thousand, and the B_2 path in Figure 27.1 is blocked.

b. i. $EP(A_1)$ $-$ 200.6 + (-50)(.4) $-$ 100;
$EP(A_2)$ $-$ 770(.3) + (-170)(.7) $-$ 112;
thus, the Bayes act is act A_2, BEP $-$ \$112 thousand and the A_1 path in Figure 27.1 is blocked.

    ii. The optimal strategy is to acquire the ski resort (A_2) and, if it succeeds, to acquire the hotel (B_1). Relevant payoffs are as follows:

Payoff	Probability
800	.3(.8) $-$ .24
650	.3(.2) $-$ .06
-170	.70
	1.00

Thus, the expected payoff for the optimal strategy is 800(.24) + 650(.06) + (-170)(.7) $-$ \$112 thousand.

Problem 5

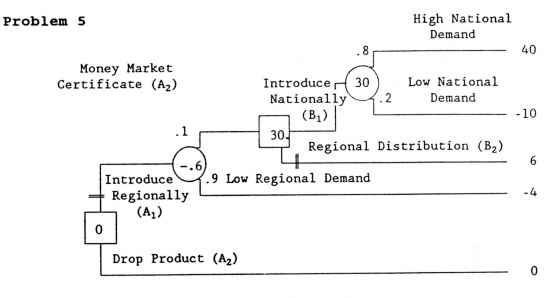

Figure 27.3

Based on the results shown in this decision tree, the optimal strategy is to drop the product from further consideration (A_2).

28

BAYESIAN DECISION MAKING WITH SAMPLE INFORMATION

A. OPTIMAL DECISION RULE FOR GIVEN SAMPLE SIZE
(Text: Chapter 28, Sections 28.1 and 28.2)

Review of Basic Concepts

A.1 The notation (5,2) refers to the decision rule: If $f \leq$ _____, select act _____ ; If $f >$ _____ select act _____, where f is the number of sample items in a sample of size n = _____ which possess the given characteristic.

A.2 Explain why the decision rule in A.1 is not appropriate for the following payoff table, where the acts refer to maintaining (A_1) or dropping (A_2) an existing brand from a firm's product line, the outcome states S_i represent different values of p, the proportion of persons in the relevant population planning to purchase this brand in the next year, and f represents the number of persons in the sample who plan to purchase this brand in the next year.

		Act	
S_i	$P(S_i)$	A_1 Maintain	A_2 Drop
S_1: .05	.6	-30	0
S_2: .10	.3	-5	0
S_3: .15	.1	25	0

A.3 For the situation in A.2, the decision rule in A.1 could be made appropriate if act _____ is selected when $f \leq 2$ and act _____ is selected when $f > 2$.

A.4 Given a decision rule (n, K), the measure of how likely it is that the decision rule will lead to the selection of A_j when the outcome state is S_i is called a(n) _____ _____ and is denoted by _____.

A.5 Consider the following payoff table, in which the acts are to buy (A_1) or not buy (A_2) a large shipment of fabricated parts and the outcome states S_i represent different values for the proportion of defective parts in the shipment.

Proportion Defective S_i	$P(S_i)$	Act	
		A_1 Buy	A_2 Do Not Buy
S_1: .02	.4	100	0
S_2: .04	.4	-20	0
S_3: .06	.2	-140	0

Using Table C-5 in your Text, determine each of the: following for the decision rule (10, 0) where f is the number of defective parts in a random sample selected from the shipment.

a. $P(A_1|S_1)$ = _____

b. $P(A_1|S_2)$ = _____

c. $P(A_1|S_3)$ = _____

A.6 For the situation in A.5, determine each of the following for the decision rule (10,0).

a. $P(A_1|S_1)$ = _____

b. $P(A_1|S_2)$ = _____

c. $P(A_1|S_3)$ = _____

A.7 For a particular decision rule in a decision problem with two acts, if $P(A_1|S_1)$ = .8171, then $P(A_2|S_1)$ = _____.

A.8 For the situation in A.5, $P(A_1|S_3)$ =.5386 and $P(A_2|S_3)$ = 1 - .5386 = .4614. Thus, EP(10,0$|S_3$) = _____. Explain what EP(10,0$|S_3$) means here. _____

A.9 For a particular decision problem with two outcome states, EP(4,2$|S_1$) = 640 and EP(4,2$|S_2$) = 350. If the prior probabilities for S_1 and S_2 are $P(S_1)$ = .7 and $P(S_2)$ = .3, then the expected payoff for decision rule (4,2), which is denoted by ____ equals ____.

A.10 For the situation in A.9, there are _____ possible decision rules (n, K) when n=4. Identify these decision rules. _____

A.11 For the situation in A.9 and A.10, the expected payoffs for the various decision rules are as follows:

K:	0	1	2	3	4	-1
EP(4, K):	452	510	553	567	531	200

Thus, the Bayes decision rule is _____. Further, the expected payoff for the Bayes decision rule, denoted by _____, equals _____. Finally, the Bayes act is act A_1 if f _____ and act A_2 if f _____, where f is the number of the sample items having the characteristic of interest and A_1 is the better act when f is small. √

»Problem 1

An auto parts restoration firm periodically has an opportunity to purchase large lots of used carburetors from a salvage company. The net profit to the restoration firm when a lot is purchased depends on the proportion of carburetors in the lot which cannot profitably be rebuilt and sold. Past experience with the salvage company indicates that in fifty percent of lots the proportion which cannot profitably be rebuilt is .20, in thirty percent of lots the proportion is .40, and in the remaining twenty percent of lots the proportion is .60. This information and the relevant payoffs (in dollars) are summarized in Table 28.1a.

Table 28.1
BAYESIAN DECISION PROBLEM FOR AUTO PARTS
RESTORATION FIRM CARBURETOR
PURCHASE DECISION
(See Problem 1)

(a) Payoff Table

i	Proportion Which Cannot Be Rebuilt S_i	$P(S_i)$	Act	
			A_1 Purchase Lot	A_2 Do Not Purchase Lot
1	.20	.5	4000	0
2	.40	.3	1000	0
3	.60	.2	-2000	0

(b)
Calculation of Expected Payoff for Decision Rule (5, 2)

S_i	Payoffs C_{i1}	C_{i2}	Act Probabilities $P(A_1\|S_i)$	$P(A_2\|S_i)$	Conditional Expected Payoff $EP(5,2\|S_i)$	$P(S_i)$	
.20	4000	0	.9421	.0579	3768.4	.5	1884.2
.40	1000	0	_____	_____	_____	.3	_____
.60	-2000	0	_____	_____	_____	.2	_____
						$EP(5,2) =$	_____

(c)
Calculation of Expected Payoff for Decision Rule (5, 3)

S_i	Payoffs C_{i1}	C_{i2}	Act Probabilities $P(A_1\|S_i)$	$P(A_2\|S_i)$	Conditional Expected Payoff $EP(5,3\|S_i)$	$P(S_i)$	
.20	4000	0	_____	_____	_____	.5	_____
.40	1000	0	_____	_____	_____	.3	_____
.60	-2000	0	_____	_____	_____	.2	_____
						$EP(5,3) =$	_____

a. i. Determine the Bayes act and BEP(0) if the decision is based
        on prior probabilities only.

    ii. What is the maximum amount the restoration firm should
        consider paying for perfect (i.e., exact) information about
        the proportion of carburetors which cannot be rebuilt in lots
        from this salvage company? _____ √

b. Since the salvage company is located in a distant city, it is
    not feasible to conduct an exhaustive examination of each lot.
    However, arrangements have been made with an independent auto
    service center to have a random sample of n = 5 carburetors from
    each lot examined. The restoration firm has decided to purchase
    a lot if 2 or fewer of the five carburetors examined cannot be
    rebuilt profitably and to not purchase a lot if 3 or more cannot
    be rebuilt profitably.

    i. Complete the computations in Table 28.1b to obtain EP(5, 2). (Note: Obtain the required act probabilities from Table C-5 in your Text.)

    ii. Interpret the meaning of EP(5, 2), which was obtained in (i.). _____

    _____

c. Use Table 28.1c to obtain the expected payoff for decision rule (5, 3). √

d. The expected payoffs (to the nearest dollar) for all possible decision rules with a sample of n = 5 carburetors are as follows:

K:	0	1	2	3	4	5	-1
EP(5, K):	675	1541	1962	1995	1927	1900	0

    i. On the basis of these expected payoffs would you recommend that the restoration firm proceed with decision rule (5,2)? Explain. _____

    _____

    _____

    ii. Complete the following table. √

X:	0	1	2	3	4	5
Bayes Act:	A_1	____	____	____	____	____

B.   DETERMINATION OF OPTIMAL SAMPLE SIZE
(Text: Chapter 28, Section 28.3)

B.1 In a particular decision problem, sample items must be subjected to an engineering stress test. The cost of setting up the necessary equipment to conduct this test is $75. In addition, there is a $30 labor and materials cost for each item tested. Therefore, the cost of a sample of n items when n > 0, which is denoted by _____, equals _____. If n=15, the total cost of the sample is _____.

B.2 Explain why no sample size should be considered for which s(n) exceeds EVPI. _____

_____

B.3 For the situation in B.1, EVPI = $700. Thus, an upper bound on the optimal sample size, which is denoted by ____ , is _____ .

B.4 For the situation in B.1, BEP(15) = $4,750. Thus, BNEP(15) = ____ .
Explain what BNEP(15) measures here. _____
_____
_____

B.5 For a particular decision problem, BEP(24) is the largest expected payoff among the Bayes decision rules for all sample sizes in the interval $0 \leq n \leq max(n)$. Does this imply that n = 24 is the optimal sample size? _____ Explain. _____
_____
_____

B.6 In a decision problem, BEP(0) = $150, BEP(10) = $650, and S(10) = $50. Obtain and interpret each of the following.

a. EVSI(10) _____
_____

b. ENGS(10) _____ ✓
_____

»Problem 2

Refer to the situation in Problem 1. Suppose the auto parts restoration firm has not yet decided on a sample size. The independent auto Service center which has agreed to examine a random sample of carburetors from each lot has indicated that they will charge a fee of $5 to pick up and return the sample of carburetors plus $15 for each carburetor examined.

a. Identify s(n) for this situation. ✓

b. Recall from Problem 1 that BEP(0) = $1,900 and BEP(5) = $1,995 (to the nearest dollar).

i. What is the Bayes net expected payoff with a sample of n = 5 carburetors? _____ ✓

ii. What is the expected value of the sample information provided by a sample of n = 5 carburetors? _____

iii. What is the expected net gain from a sample of n = 5 carburetors? _____ √

c. In Problem 1(a.ii.) we found that EPPI = $2,300 and EVPI = $400.

i. Given s(n) = 5 + 15n for n > 0, determine the upper bound on the optimal sample size. _____ √

ii. In fact, the independent auto service center has indicated that it is willing to examine no more than 10 carburetors from any lot. Since n = 10 is less than max(n), is n = 10 the best sample size among those available to the restoration firm (i.e., n = 0 through n = 10)? _____ Explain. _____

iii. Table 28.2 shows the Bayes rule and BEP(n) for samples of size n = 0 through n = 10. Complete Table 28.2 by determining s(n) and BNEP(n) for samples of size n = 7 through n = 10. √

Table 28.2
BAYES DECISION RULES FOR n=0 THROUGH n=10
FOR AUTO PARTS RESTORATION FIRM DECISION PROBLEM
(See Problem 2(c.iii.))

Bayes Rule				
n	K	BEP(n)	s(n)	BNEP(n)
0	-	1,900	0	1,900
1	1	1,900	20	1,880
2	1	1,916	35	1,881
3	2	1,951	50	1,901
4	2	1,982	65	1,917
5	3	1,995	80	1,915
6	3	2,039	95	1,935
7	3	2,031		
8	4	2,065		
9	4	2,075		
10	5	2,091		

iv. What is the best sample size among those available to the restoration firm? _____ Explain. _____

√

C. DETERMINING BAYES ACT BY REVISION OF PRIOR PROBABILITIES

(Text: Chapter 28, Section 28.4)

Review of Basic Concepts

C.1 When prior probabilities have been revised with Bayes' theorem on the basis of observed sample results, the revised probabilities are called _____ probabilities.

C.2 In a decision problem with two outcome states, prior probabilities $P(S_1) = .25$ and $P(S_2) = .75$ have been assigned. A random sample of $n = 10$ items is examined, of which three possess the characteristic of interest. This sample result is denoted by _____. In general, if j sample items possess the characteristic of interest, then this sample result is denoted by _____.

C.3 For the situation in C.2, it is determined from the sampling distribution of X_j, that the probability of obtaining the sample result X_3 is .2503 if outcome state S_1 prevails and .1172 if outcome state S_2 prevails. These probabilities are called _____ probabilities and are denoted by ____ , ____ , and ____ respectively.

C.4 For the situation in C.2 and C.3, $P(S_1|X_3)$ is called the _____ probability of outcome state _____ given the sample result _____. Using Bayes' theorem, $P(S_1|X_3) = $ _____. Similarly, $P(S_2|X_3) = $ _____.

C.5 The procedure in which prior probabilities are revised with Bayes' theorem is called the _____ form of Bayesian decision analysis in contrast to the procedure in which prior probabilities are applied to conditional expected payoffs, which is called the _____ form.

C.6 Prior probabilities are introduced last in the analysis with the _____ form of Bayesian analysis.

C.7 Using the normal form of Bayesian analysis, the Bayes decision rule for a sample of size $n = 8$ was found to be: If $f \leq 6$, select A_1; if $f > 6$, select A_2. Suppose a random sample of $n = 8$ items is selected and the sample result $f4$ is obtained. If Bayes' theorem is used to revise the prior probabilities, act _____ will yield the maximum expected payoff under the posterior probabilities. ✓

»Problem 3

A manufacturing company holds a patent for a newly developed industrial tool. It is in the process of deciding whether to sell the patent to another firm (A_1) for $300 thousand or retain the patent and market the tool itself (A_2). In the latter case, the payoff depends upon the proportion of industrial firms in the market area that will purchase the product (S_i). Preliminary cost and market forecasts were used to develop the payoff table which appears in Table 28.3a.

Table 28.3
BAYESIAN DECISION PROBLEM FOR
MANUFACTURING COMPANY PATENT DECISION
(See Problem 3)

(a)
Payoff Table ($ thousands)

| | Proportion of Firms That Will Purchase Product | | Act | |
| | | | A_1 | A_2 |
i	S_i	$P(S_i)$	Sell Patent	Market Product
1	.05	.20	300	-400
2	.10	.30	300	-100
3	.15	.20	300	250
4	.20	.15	300	650
5	.25	.10	300	1100
6	.30	.05	300	1600

(b)
Revision of Prior Probabilities for Sample Result X_1

i	Outcome State S_i	Prior Probability $P(S_i)$	Conditional Probability $P(X_1\|S_i)$	Joint Probability $P(S_i \cap X_1)$	Posterior Probability $P(S_i\|X_1)$
1	.05	.20	.3774	.07548	_____
2	.10	.30	.2702	.08106	_____
3	.15	.20	_____	_____	_____
4	.20	.15	_____	_____	_____
5	.25	.10	_____	_____	_____
6	.30	.05	_____	_____	_____
	Total	1.00			

The sales manager of the company reported that he has just completed a survey of n = 20 randomly selected industrial firms in the market area with somewhat disappointing results. Just one of the twenty firms surveyed (X_1) indicated that it would purchase this new product.

a. Without performing any calculations, indicate how the pattern of posterior probabilities $P(S_i|X_1)$ might compare with the Prior Probabilities in Table 28.3a. _____

_____

_____ √

b. i. Using Table C-5 in your Text, obtain the conditional probabilities $P(X_1|S_i)$ for the sample result X_1 given the outcome states S_i and insert these in Table 28.3b. √

   ii. Complete the computations in Table 28.3b to obtain the posterior probabilities $P(S_i|X_1)$. √

c. i. Determine the expected payoff for each act under the posterior probabilities. $EP(A_1|X_1)$ = _____ ;
   $EP(A_2|X_1)$ = _____ .

   ii. Which act is the Bayes act? _____ Explain. _____

   _____

   _____ √

Answers to Chapter Reviews and Problems

A.1 If $f \le 2$, select act A_1; If $f > 2$, select act A_2; n = 5

A.2 The decision rule in A.1 leads to the selection of act A_1 when p is relatively small and A_2 when p is relatively large. In this case, however, act A_2 offers a higher payoff when p is relatively small and A_1 offers a higher payoff when p is relatively large.

A.3 A_2; A_1

A.4 act probability; $P(A_j|S_i)$

A.5 a. $P(A_1|S_1)$ = .8171
    b. $P(A_1|S_2)$ = .6648
    c. $P(A_1|S_3)$ = .5386

A.6 a. $P(A_1|S_1) = .9838$
     b. $P(A_1|S_2) = .9418$
     c. $P(A_1|S_3) = .8824$

A.7 $P(A_2|S_1) = 1 - .8171 = .1829$

A.8 $EP(10, 0|S_3) = -140(.5386) + 0(.4614) = -75.404$; $EP(10,0|S_3)$ is
the conditional expected payoff for decision rule (10, 0) when the
outcome state is S_3 (i.e., when the proportion of defective parts
in the shipment is .06).

A.9 $EP(4,2) = 640(.7) + 350(.3) = 553$

A.10 six; (4, 0), (4, 1), (4, 2), (4, 3), (4, 4), (4, -1)

A.11 (4, 3); BEP(4) = 567; $f \leq 3$; $f > 3$

Problem 1

   a. i. $EP(A_1) = 4,000(.5) + 1,000(.3) + (-2,000)(.2) = 1,900$;
         $EP(A_2) = 0(.5) + 0(.3) + 0(.2) = 0$;
         Thus, the Bayes act is act A_1 and BEP(0) = \$1,900.

     ii. EPPI = $4,000(.5) + 1,000(.3) + 0(.2) = 2,300$;
         Thus, EVPI = 2,300 - 1,900 = \$400.

   b. i.

			Conditional Expected Payoff		
	Act Probabilities				
i	$P(A_1|S_i)$	$P(A_2|S_i)$	$EP(5,2|S_i)$	$P(S_i)$	
1	.9421	.0579	3768.4	.5	1884.2
2	.6826	.3174	682.6	.3	204.8
3	.3174	.6826	-634.8	.2	-127.0
				$EP(5,2) =$	1,962.0

     ii. If the decision rule (5, 2) is used repeatedly with many
        purchase opportunities under the given prior probabilities
        and payoffs, the average payoff per purchase opportunity in
        the long run is \$1,962.

c.

i	Act Probabilities $P(a_1\|S_i)$	$P(A_2\|S_i)$	Conditional Expected Payoff $EP(5,3\|S_i)$	$P(S_i)$	
1	.9933	.0067	3973.2	.5	1986.6
2	.9130	.0870	913.0	.3	273.9
3	.6630	.3370	-1326.0	.2	-265.2
			$EP(5,3)$ =		1995.3

d. i. No; decision rule (5, 3) has the largest expected payoff.
Hence, (5, 3) is the Bayes decision rule and BEP(5) = 1,995.

ii.

X:	0	1	2	3	4	5
Bayes act:	A_1	A_1	A_1	A_1	A_2	A_2

B.1 $s(n) = 75 + 30n$; $s(15) = 75 + 30(15) = \$525$

B.2 EVPI is the upper limit on the expected value of any additional
information. Therefore, if the cost of sampling n items exceeds
this amount the cost of obtaining this sample information is
greater than its expected value.

B.3 $\max(n) = (700-75)/30 = 20.83$, so the optimal sample size is not
larger than 20.

B.4 BNEP(15) = 4750 - 525 = 4225; Thus, the expected payoff for the
Bayes decision rule for n = 15 minus the sampling cost or, simply,
the net expected payoff for the Bayes decision rule for n = 15 is
\$4,225.

B.5 No; the optimal sample size is that which maximizes BNEP(n), which
may or may not be the same sample size as that which maximizes
BEP(n).

B.6 a. EVSI(10) = 650 - 150 = 500; Thus, the maximum amount that should
be paid for a sample of size n = 10 is \$500.

   b. BNEP(10) = 650 - 50 = 600; ENGS(10) = 600 - 150 = 450; Thus, the
expected gain with a sample of size n = 10 versus prior
information only is \$450 after deducting the cost of sampling.

Problem 2

a. s(n) = 5+15n for n > 0; s(0) = 0

b. i. s(5) = 5 + 15(5) = 80; Thus, BNEP(5) = 1995 - 80 = $1,915.

 ii. EVSI(5) = 1995 - 1900 = $95

 iii. ENGS(5) = 1915 - 1900 = $15

c. i. max(n) = (400-5)/15 = 26.33, so the optimal sample size is not larger than 26.

 ii. Not necessarily. We need to find the sample size which maximizes BNEP(n) among the available sample sizes n = 0,...,10.

 iii.

n	Bayes Rule K	BEP(n)	s(n)	BNEP(n)
7	3	2,031	110	1,921
8	4	2,065	125	1,940
9	4	2,075	140	1,935
10	5	2,091	155	1,936

 iv. Since BNEP(8) is the maximum Bayes net expected payoff among the sample size n = 0 through n = 10, the sample size n = 8 is the best sample size among those available.

C.1 posterior

C.2 X_3; X_j

C.3 conditional; $P(X_3|S_1)$ = .2503; $P(X_3|S_2)$ = .1172

C.4 posterior; S_1; X_3; $P(S_1|X_3)$ = [.25(.2503)]/[.25(.2503)+.75(.1172)] = .416; $P(S_2|X_3)$ = .584

C.5 extensive; normal

C.6 normal

C.7 A_1

Problem 3

a. Since the sample proportion was small ($\bar{p} = 1/20 = .05$), the posterior probabilities for outcome states representing small values of p should be larger than the corresponding prior probabilities while the posterior probabilities for outcome states representing large values of p should be smaller than the corresponding prior probabilities.

b. i. $P(X_1|S_3) = .1368$; $P(X_1|S_4) = .0576$; $P(X_1|S_5) = .0211$; $P(X_1|S_6) = .0068$.

ii.

i	S_i	$P(S_i)$	$P(X_1|S_i)$	$P(S_i \cap X_1)$	$P(S_i|X_1)$
1	.05	.20	.3774	.07548	.3871
2	.10	.30	.2702	.08106	.4157
3	.15	.20	.1368	.02736	.1403
4	.20	.15	.0576	.00864	.0443
5	.25	.10	.0211	.00211	.0108
6	.30	.05	.0068	.00034	.0018

$$P(X_1) = .19499 \qquad 1.0000$$

c. i. $EP(A_1|X_1) = 300(.3871) + 300(.4157) + 300(.1403) + 300(.0443) + 300(.0108) + 300(.0018) = \$300$ thousand.

$EP(A_2|X_1) = -400(.3871) + -100(.4157) + 250(.1403) + 650(.0443) + 1100(.0108) + 1600(.0018) = \$118$ thousand.

ii. Since act A_1 has the larger expected payoff, the Bayes act is A_1, to sell the patent.

NOTES

NOTES